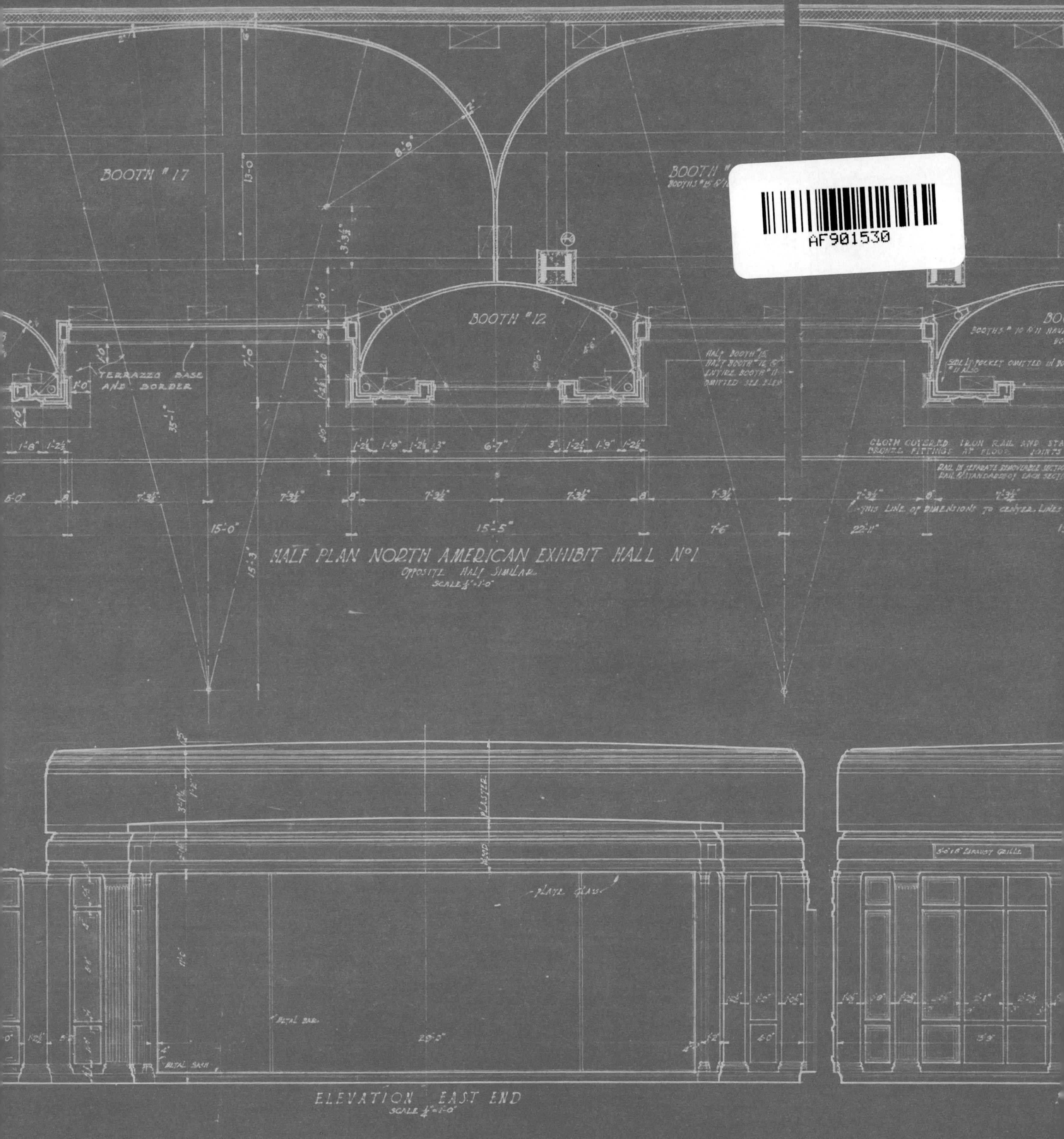
BOOTH #17
BOOTH #12
TERRAZZO BASE AND BORDER
HALF PLAN NORTH AMERICAN EXHIBIT HALL No 1
OPPOSITE HALF SIMILAR
SCALE 1/4"=1'-0"
CLOTH COVERED IRON RAIL AND STA
BRONZE FITTINGS AT FLOOR JOINTS
THIS LINE OF DIMENSIONS TO CENTER LINES
PLASTER
PLATE GLASS
METAL BAR
METAL SASH
ELEVATION EAST END
SCALE 1/4"=1'-0"

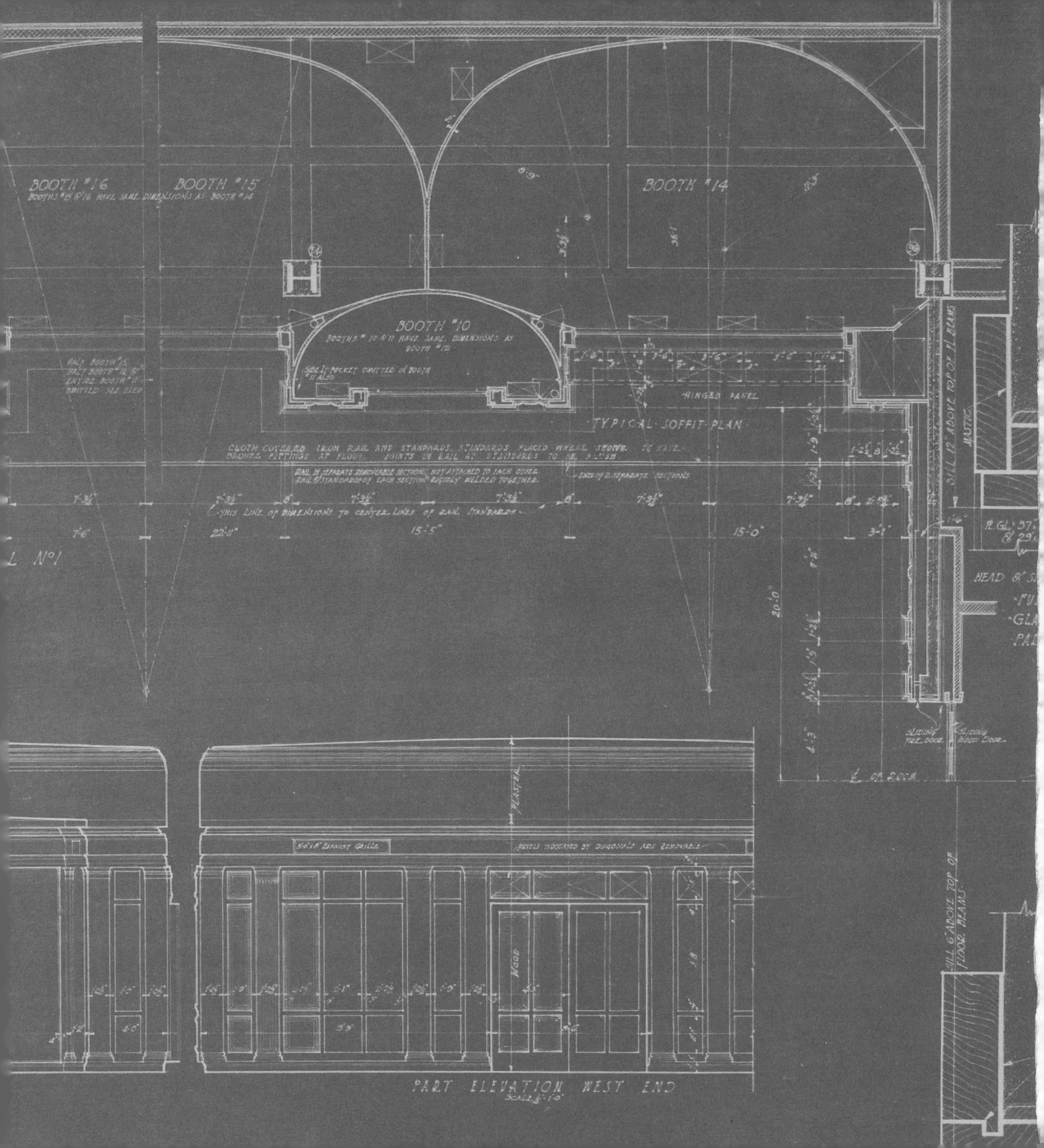
BOOTH #16
BOOTH #15
BOOTHS #15 & 16 HAVE SAME DIMENSIONS AS BOOTH #14
BOOTH #14
BOOTH #10
HINGED PANEL
TYPICAL SOFFIT PLAN
CLOTH COVERED IRON RAIL AND STANDARDS. STANDARDS PLACED WHERE SHOWN. TO HAVE BRONZE FITTINGS AT FLOOR. JOINTS IN RAIL AT STANDARDS TO BE FLUSH.
THIS LINE OF DIMENSIONS TO CENTER LINES OF RAIL STANDARDS
SILL 13" ABOVE TOP OF FL. BEAMS
MASTIC
HEAD & S
GLA
PLASTER
EXHAUST GRILLE
PANELS INDICATED BY DIAGONALS ARE REMOVABLE
WOOD
SILL 6" ABOVE TOP OF FLOOR BEAMS
PART ELEVATION WEST END

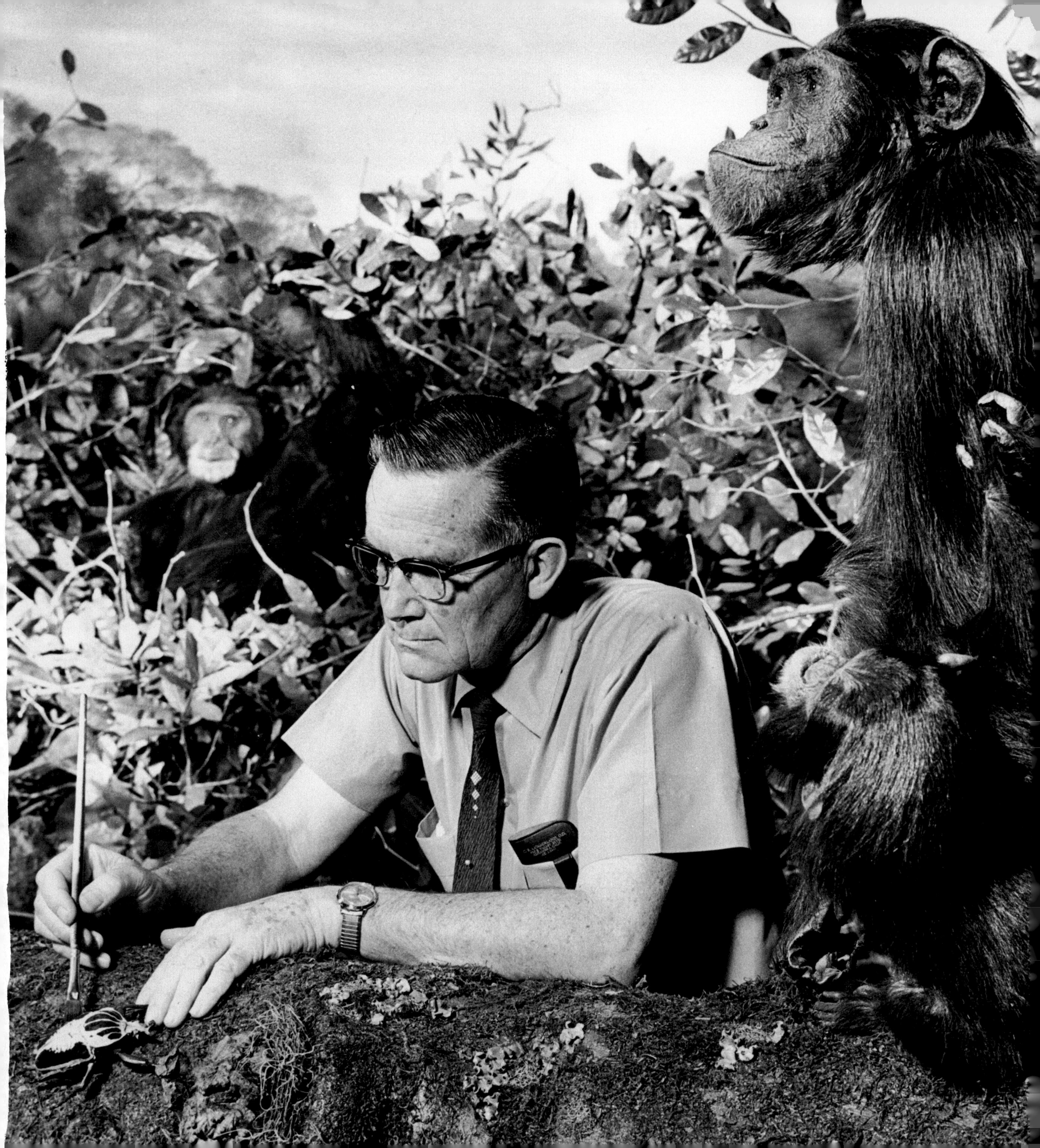

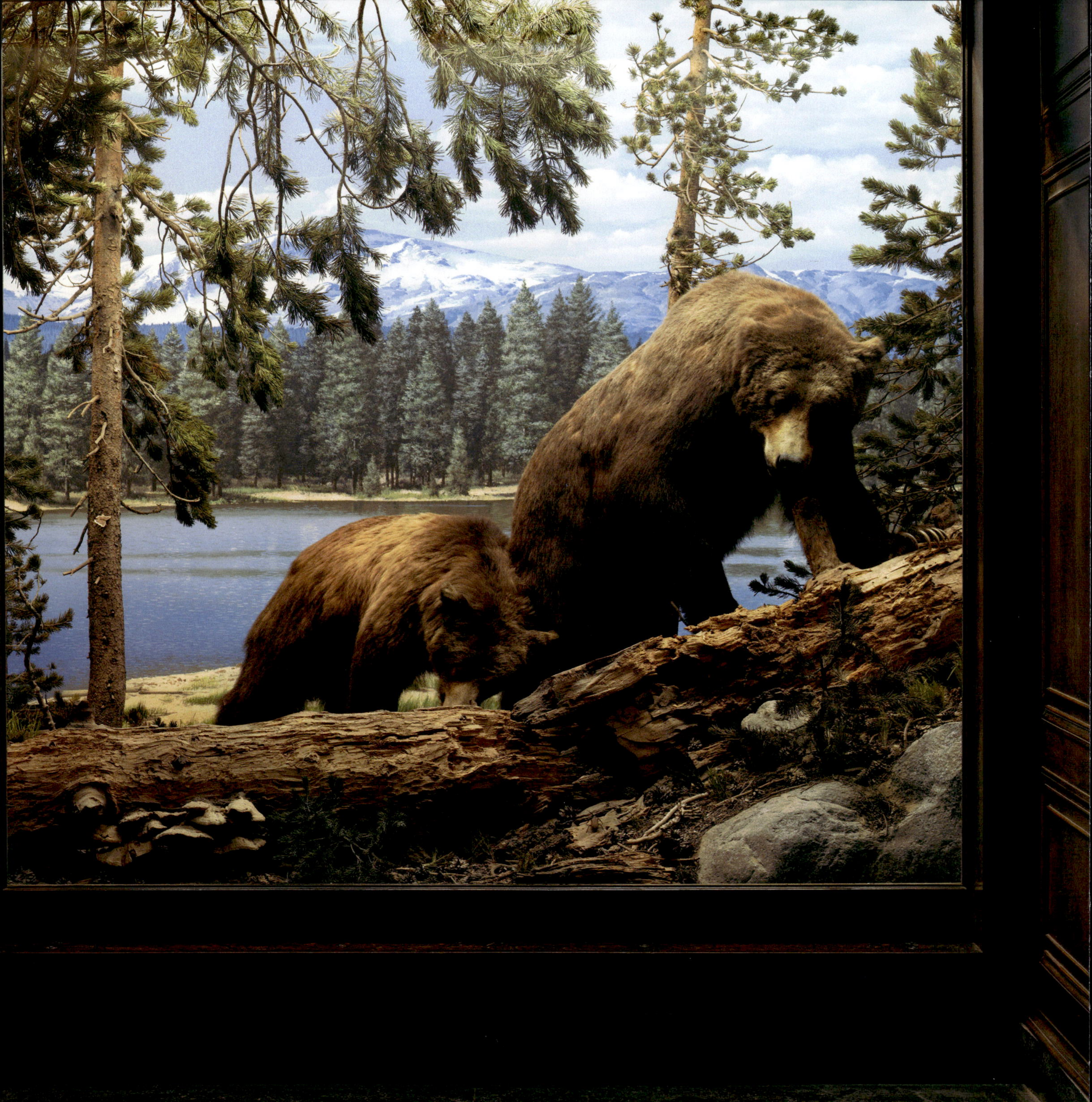

Fabricating Wilderness

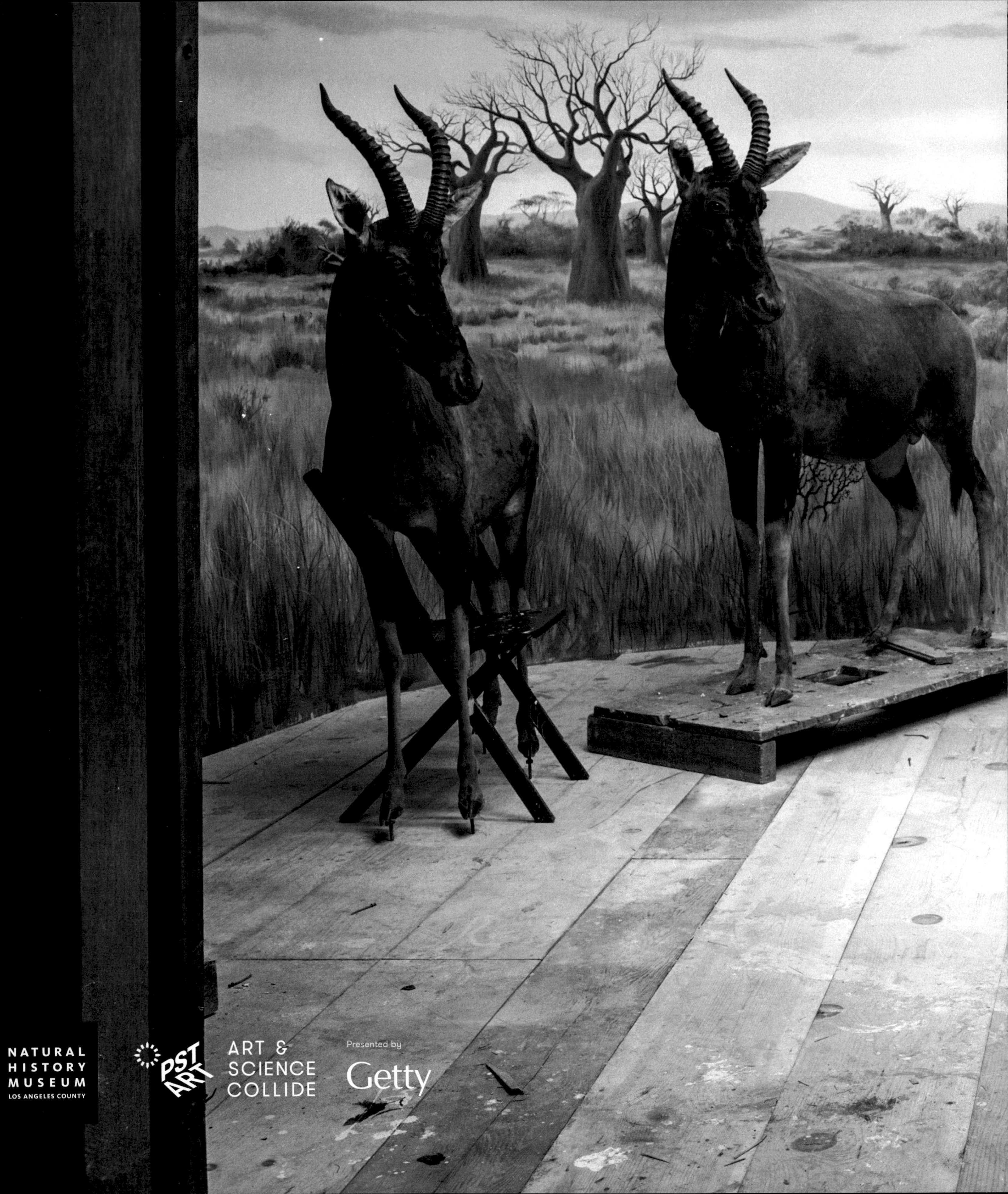
NATURAL HISTORY MUSEUM
LOS ANGELES COUNTY
PST ART
ART & SCIENCE COLLIDE
Presented by
Getty

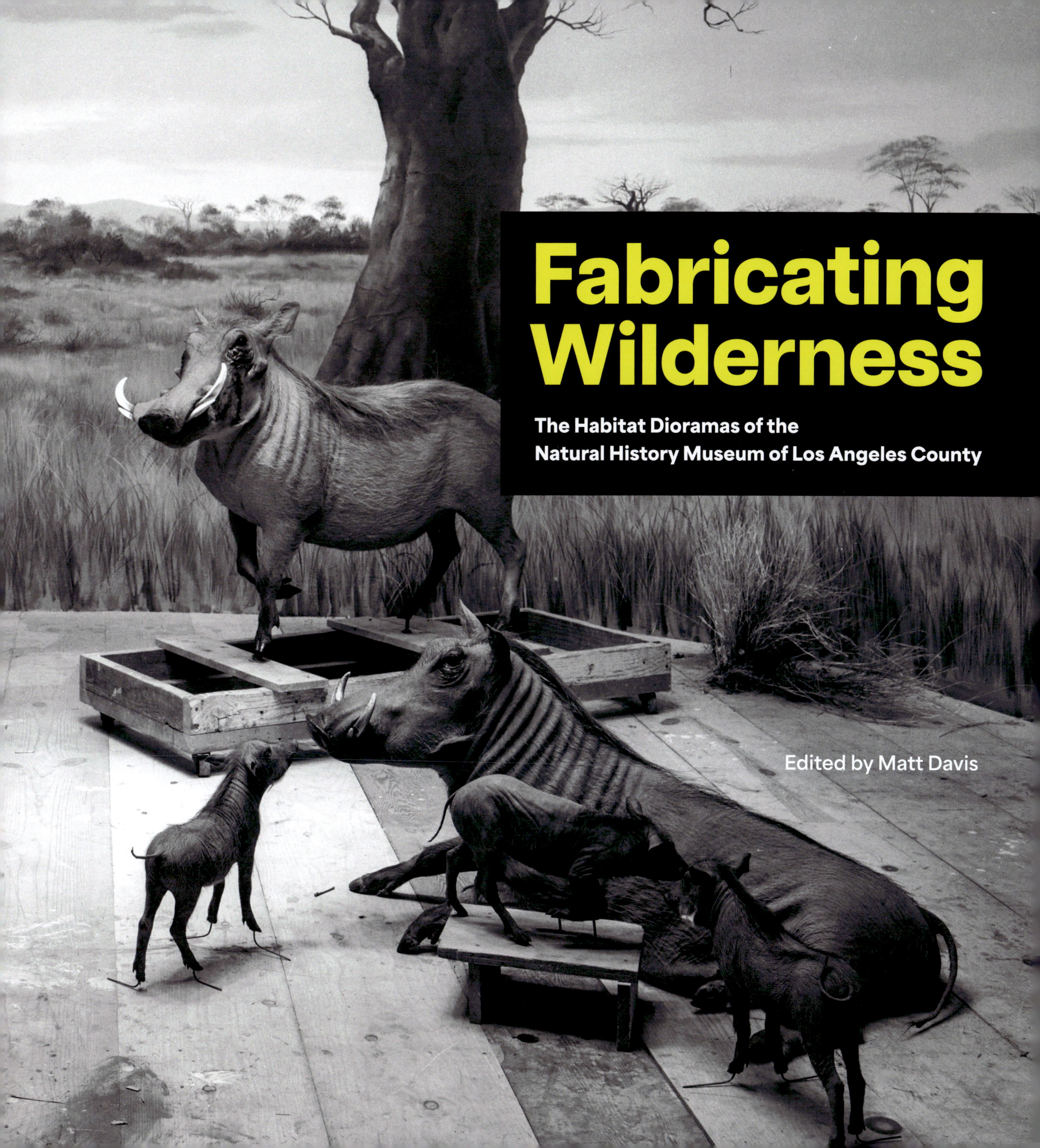

Fabricating Wilderness

The Habitat Dioramas of the Natural History Museum of Los Angeles County

Edited by Matt Davis

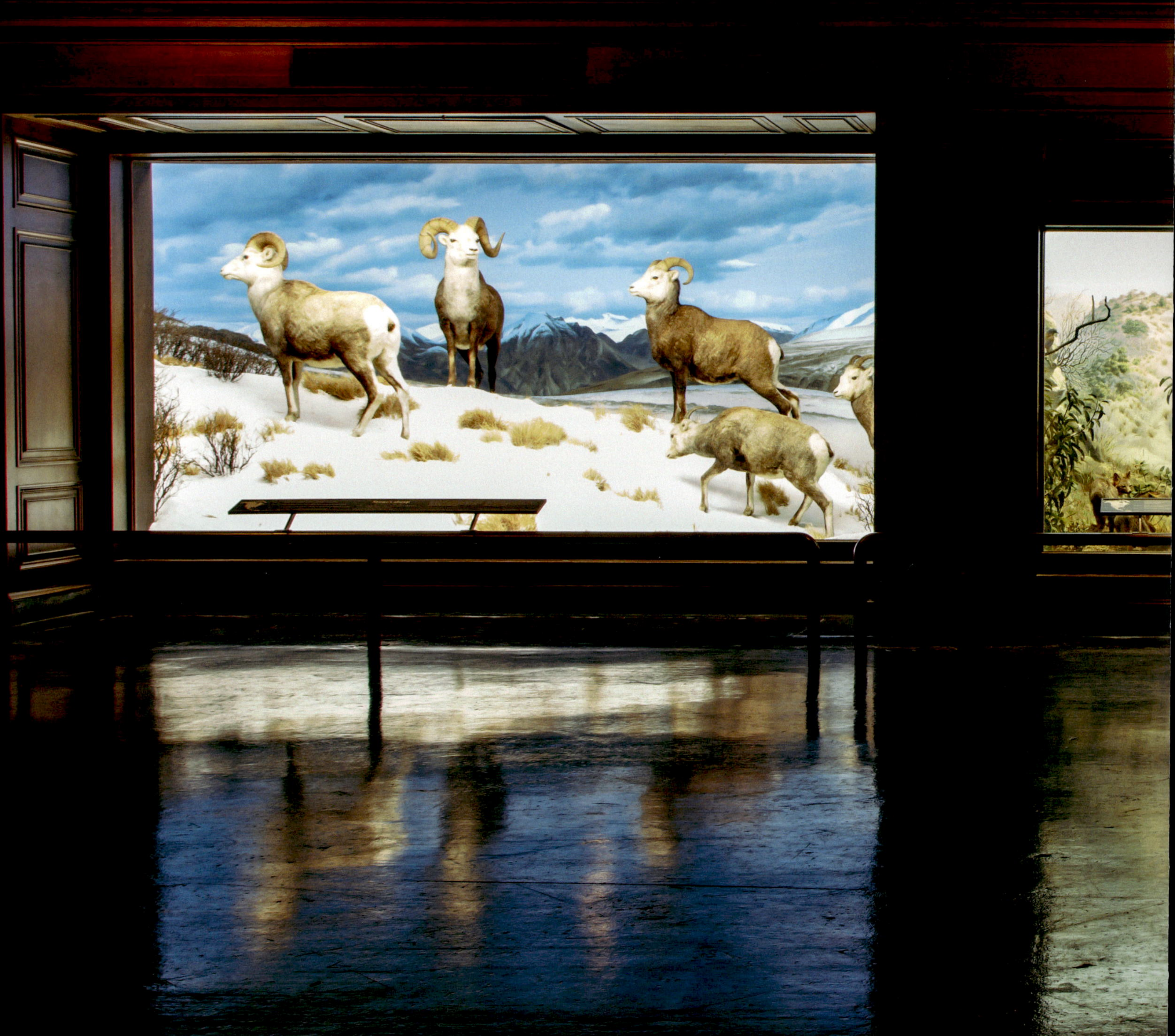

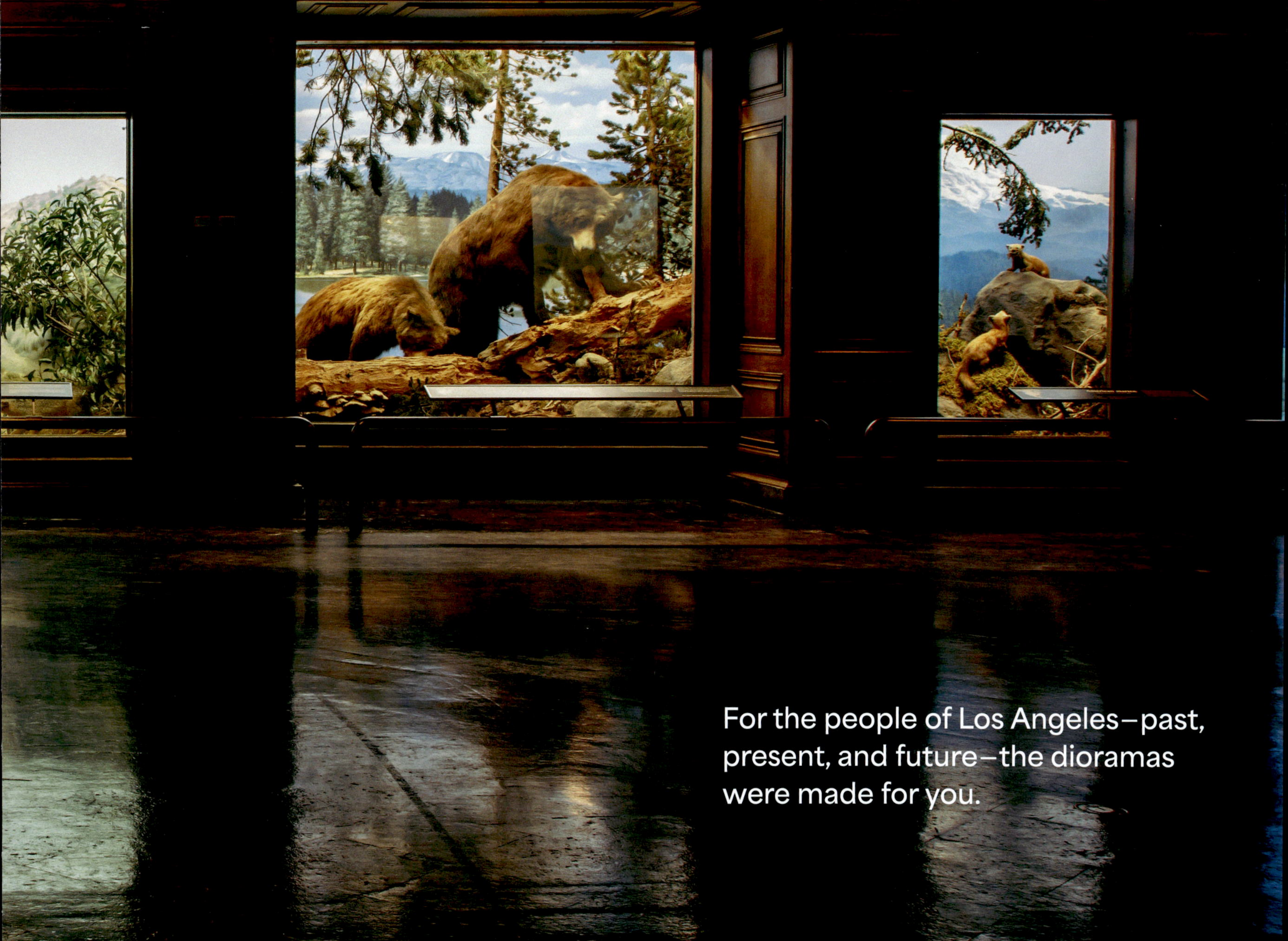

For the people of Los Angeles—past, present, and future—the dioramas were made for you.

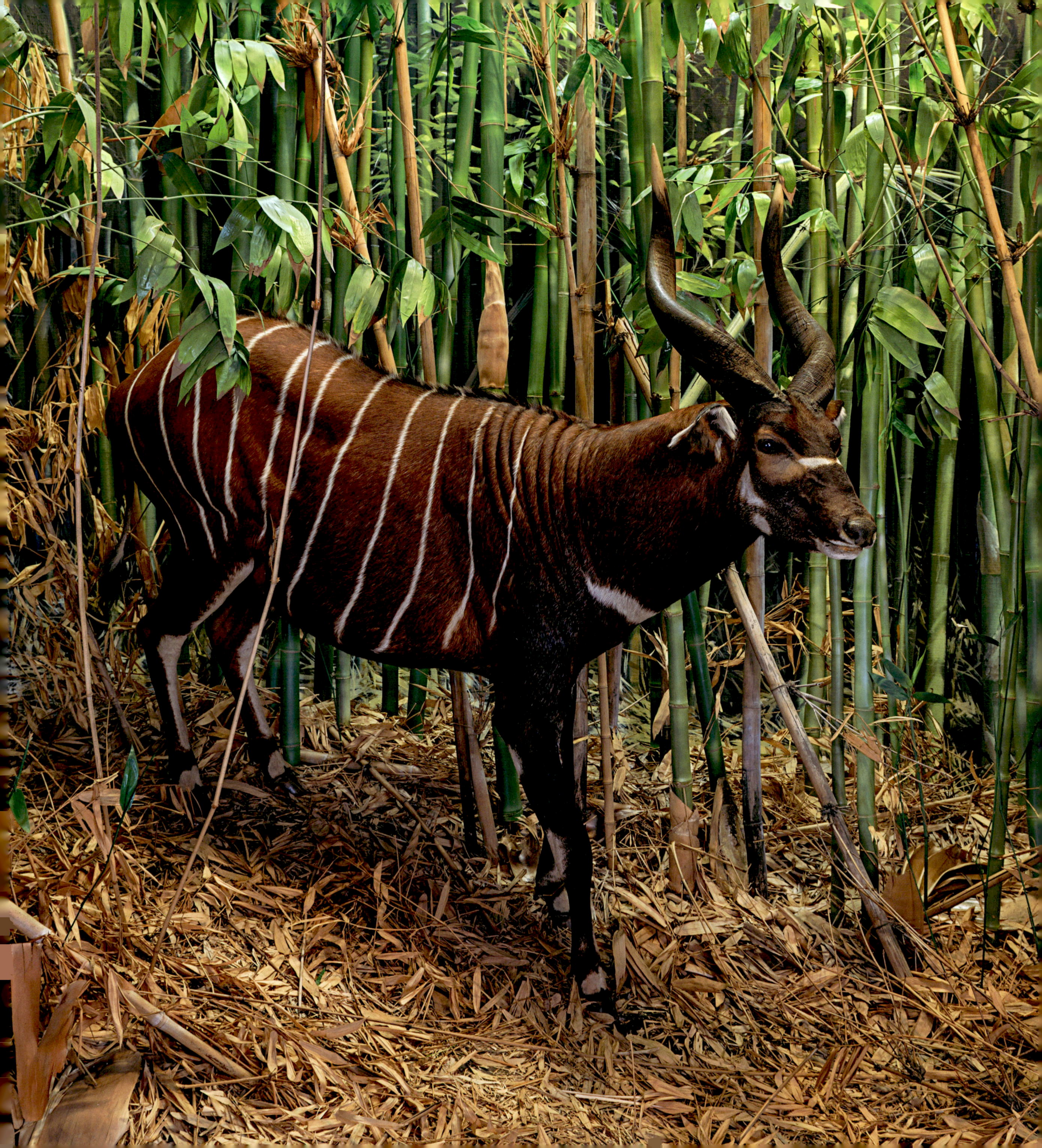

Contents

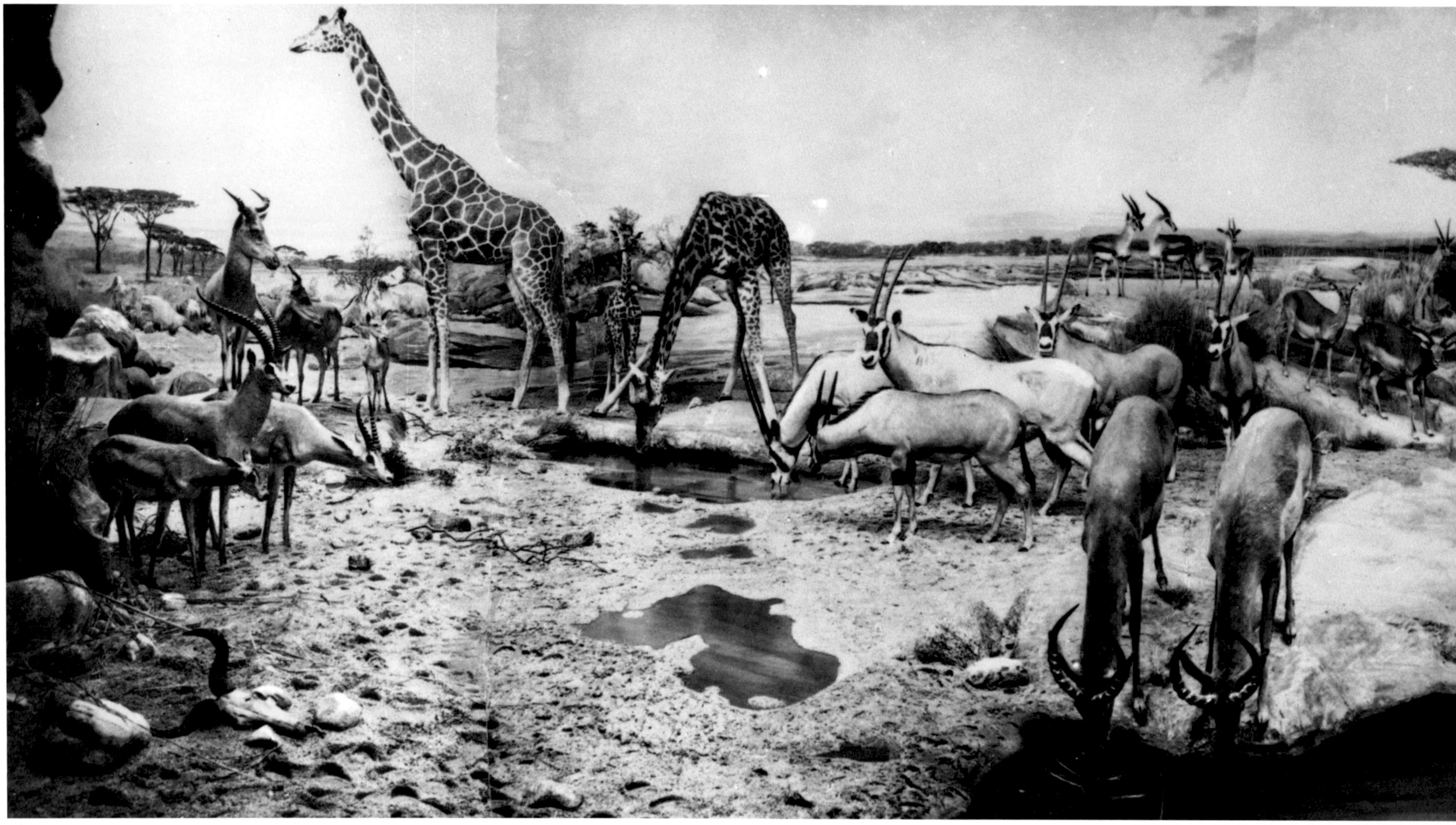

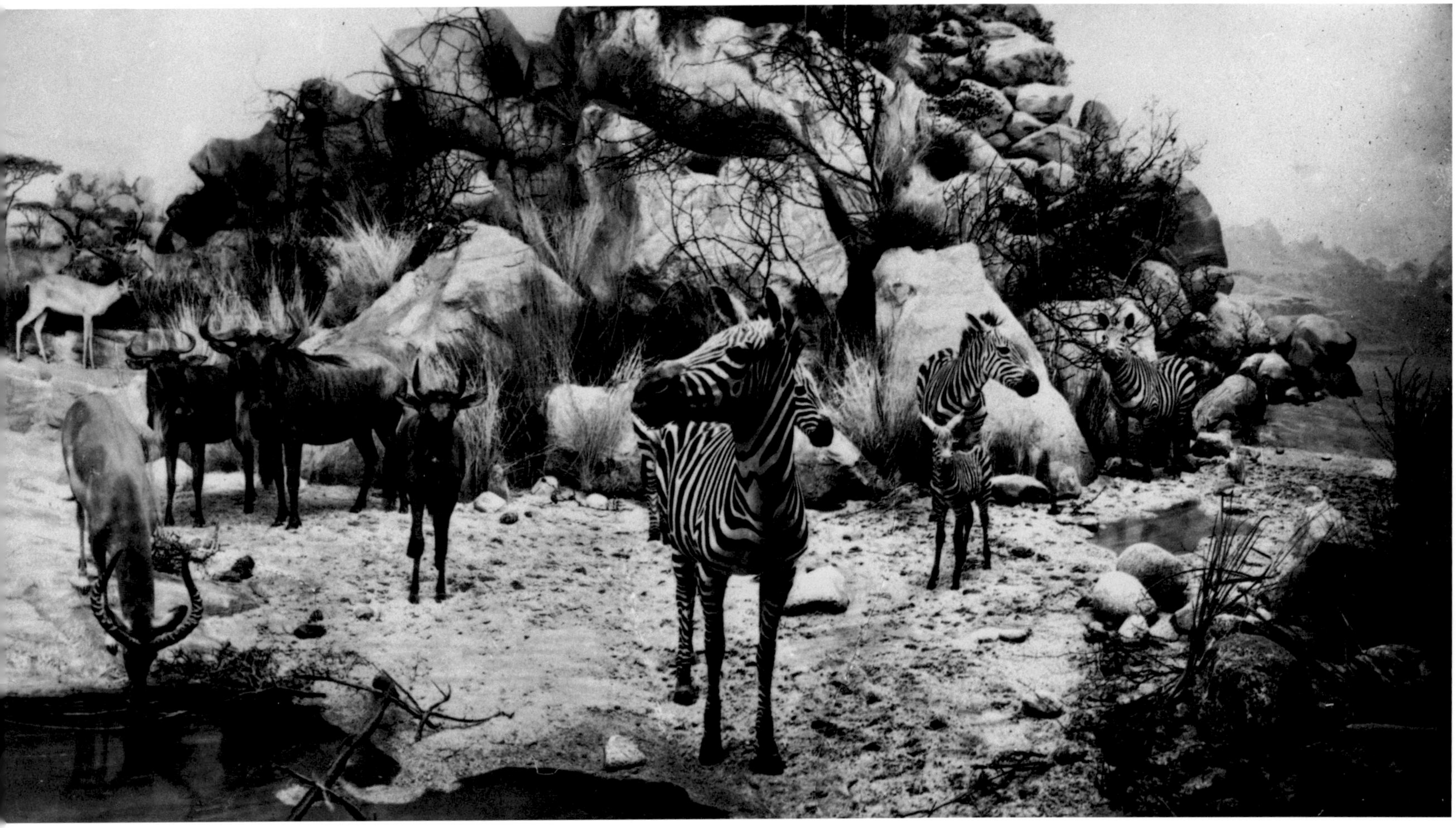

↑ The first *Watering Hole* group at the Natural History Museum of Los Angeles County (NHM) was the largest diorama in the world when it opened in 1930. When the Field Museum in Chicago announced in 1933 that its own newly constructed *Waterhole* group was the largest, NHM Director William Bryan could not resist writing to the Field Museum to correct the misunderstanding. He cheekily pointed out that NHM's was 78 feet 8 inches wide, 45 feet 1 inch deep, and 34 feet high (24 × 13.7 × 10.4 m), and thus nearly four times larger than the Field Museum's. Still, he congratulated them on "the completion of what is undoubtedly a splendid achievement (for Chicago)."[1]

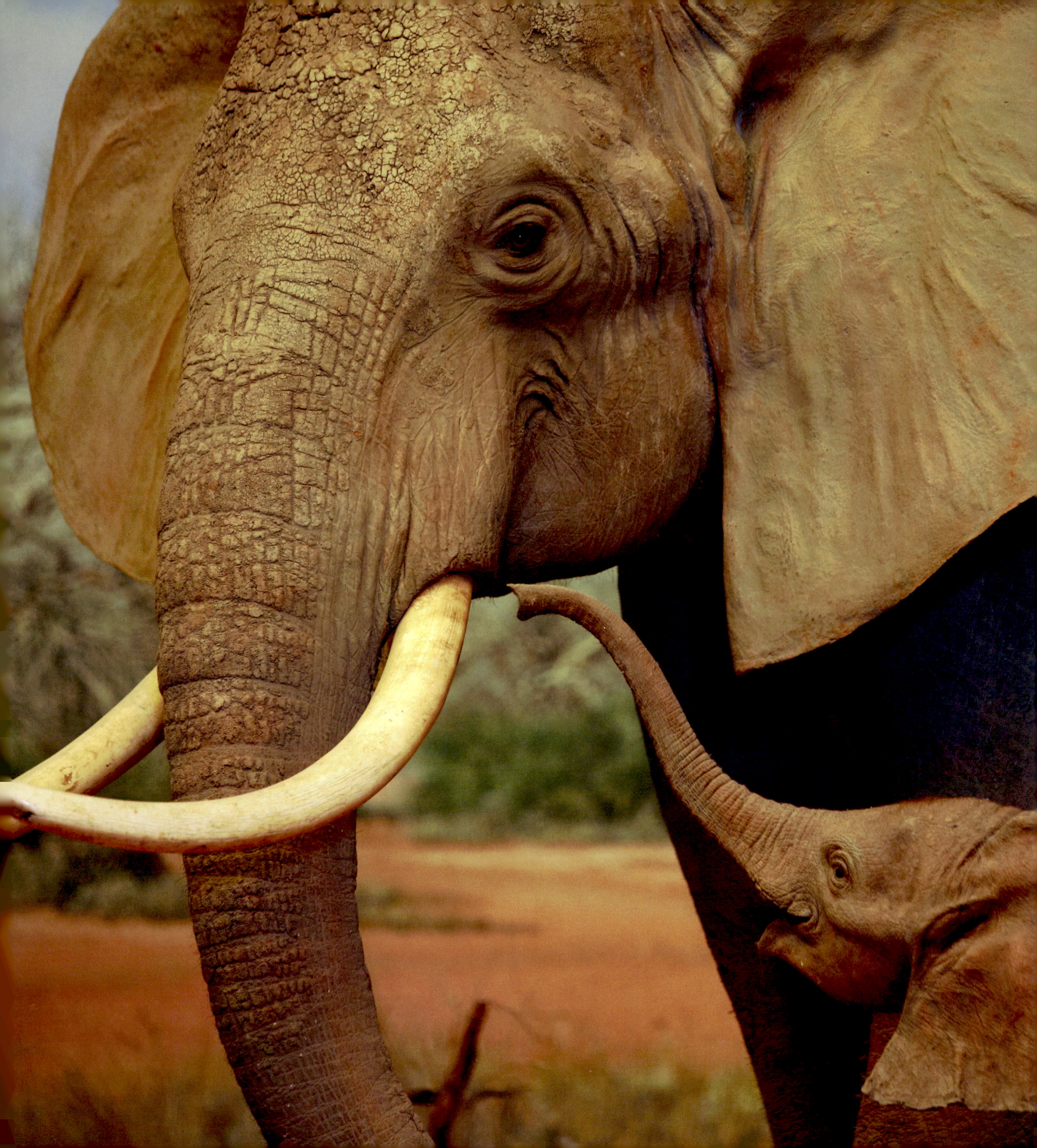

Foreword

To inspire wonder, discovery, and responsibility for our natural and cultural worlds.

—The vision of the Natural History Museum of Los Angeles County

More than 100 years ago, my predecessor Frank Daggett, the inaugural director of the Los Angeles County Museum of History, Science and Art, wrote to the County Board of Supervisors to notify them that specimens from our first major expedition had arrived safely back at the museum from Alaska. Contemplating the antlers and hides of caribou and moose then stacked outside the building, he wrote, "This material, when mounted, will always be considered the beginning of what every citizen in Los Angeles County hopes sometime to see at Exposition Park—the largest and best collection of mammal groups in the United States."[2]

As the current president and director of what is now the Natural History Museum of Los Angeles County (NHM), it's my privilege to share the story of the museum's diorama halls, which Daggett could only envision. Told by artists and researchers from both inside and outside NHM, it's the story of how painters, model makers, taxidermists, and scientists worked together to create a revolution in exhibitions that would transform museums

← A baby elephant reaches toward its mother in the *Watering Hole* diorama at the Natural History Museum of Los Angeles County.

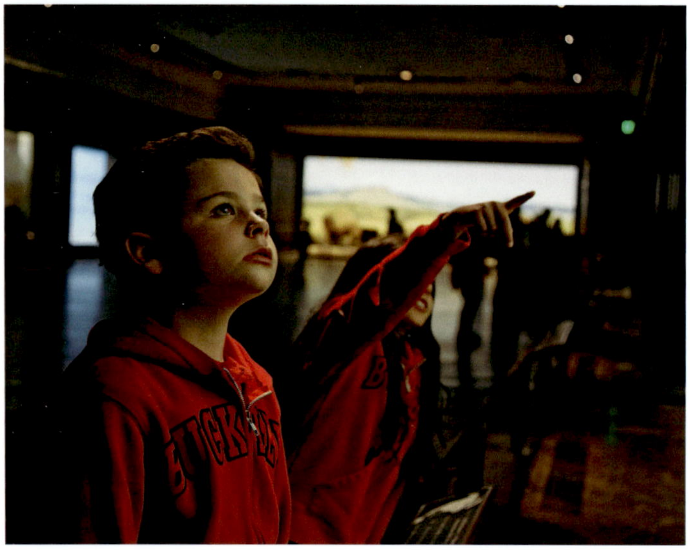

← For many patrons, their fondest memory of NHM is visiting the diorama halls as children on a school field trip (2017).

→ Senior Curator of Habitats Melville Lincoln welcomes "Teddy," the world's largest Kodiak bear, into the museum to join the other taxidermied animals in the diorama halls (1953).

across the country. NHM opened in 1913, during a time of overwhelming environmental destruction. Alarmed by the rapidly growing number of extinctions and regional exterminations (passenger pigeon, California grizzly bear, heath hen, and on and on), museum scientists and artists rushed to preserve as many species, and even entire habitats, as possible inside dioramas before it was too late.

Dioramas did more than preserve re-creations. Before dioramas became so popular, museums predominantly featured displays of skeletal remains. Dioramas ushered in a revolutionary approach, merging art and drama to convey scientific knowledge. In doing so, they made museums accessible to the general public, not just the educated elite. Unlike the sterile cases of bones, dioramas gave everyday visitors something they could identify with and fight to save—majestic animals, such as bison, in habitats that could soon be lost forever. Our habitat group dioramas were an important part of America's early conservation movement, which saved many of the species inside them from extinction. Daggett would have been proud.

Yet as dioramas stay still, the world changes around them. NHM would never support an expedition today to hunt lions or other threatened animals to put on display, and nor would most Angelenos. Like other museums working toward decolonization, we are rethinking what it means to have collections of animals obtained decades ago from colonized lands. Or to show idealized scenes of virgin wilderness in dioramas, erasing the peoples who have been living in those habitats for generations.

The original purpose of dioramas, which was to preserve entire habitats and inspire their conservation, hasn't changed, though. In fact, with nearly a quarter of the world's mammals facing extinction, dioramas are perhaps more significant now than ever—and they can be effective. Thanks to conservation measures partially inspired by dioramas, animals like bison and tule elk are more common today than a century ago. Notably, NHM never stopped building and updating our dioramas. We've installed newer habitat groups featuring specimens not shot in the wild but instead donated by conservation centers and zoos. They showcase threatened species like the okapi and the snow leopard that grow rarer every day.

In 2024, as part of Getty's PST ART: *Art & Science Collide* initiative, NHM reopened a long-closed diorama hall. This reopening serves a dual purpose: to critically reexamine the legacy of dioramas and to reaffirm our commitment to this century-old art form. Our dioramas have been enjoyed by four generations of Angelenos, making them a significant part of our cultural heritage. They inspire wonder with their artistry and immersive realism. They encourage discovery and open-ended exploration with each animal hidden in their foliage. And they imbue a deep sense of responsibility for our natural and cultural worlds by bringing wilderness into the heart of Los Angeles.

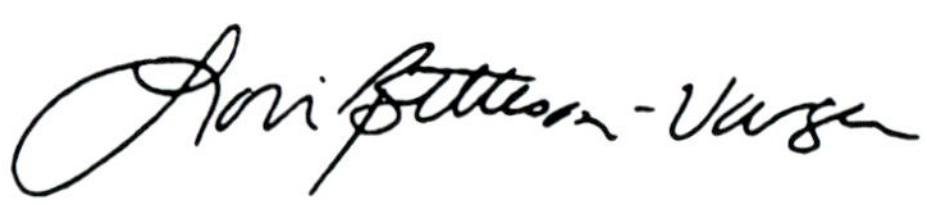

Lori Bettison-Varga

President and Director
Natural History Museums of Los Angeles County

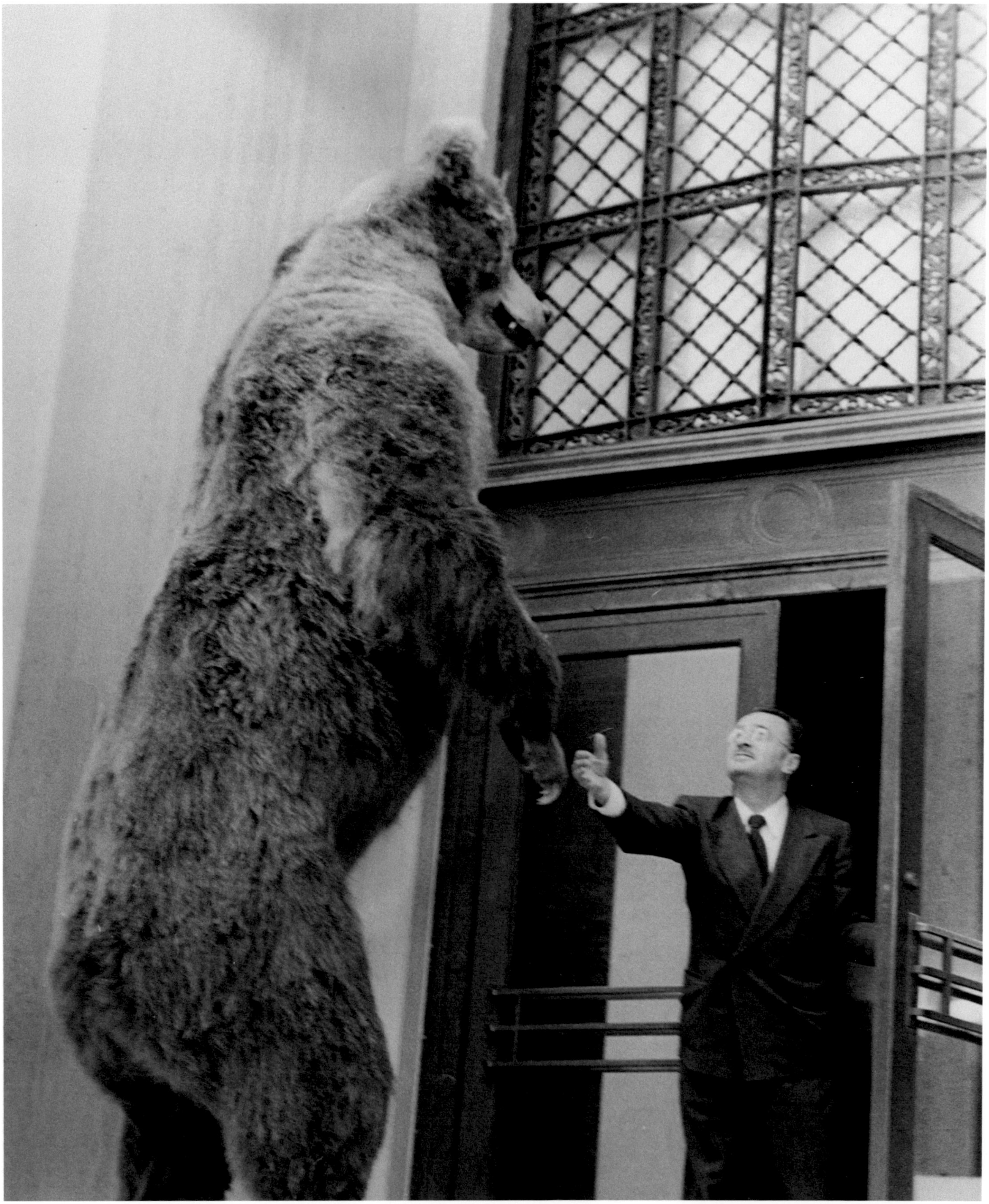

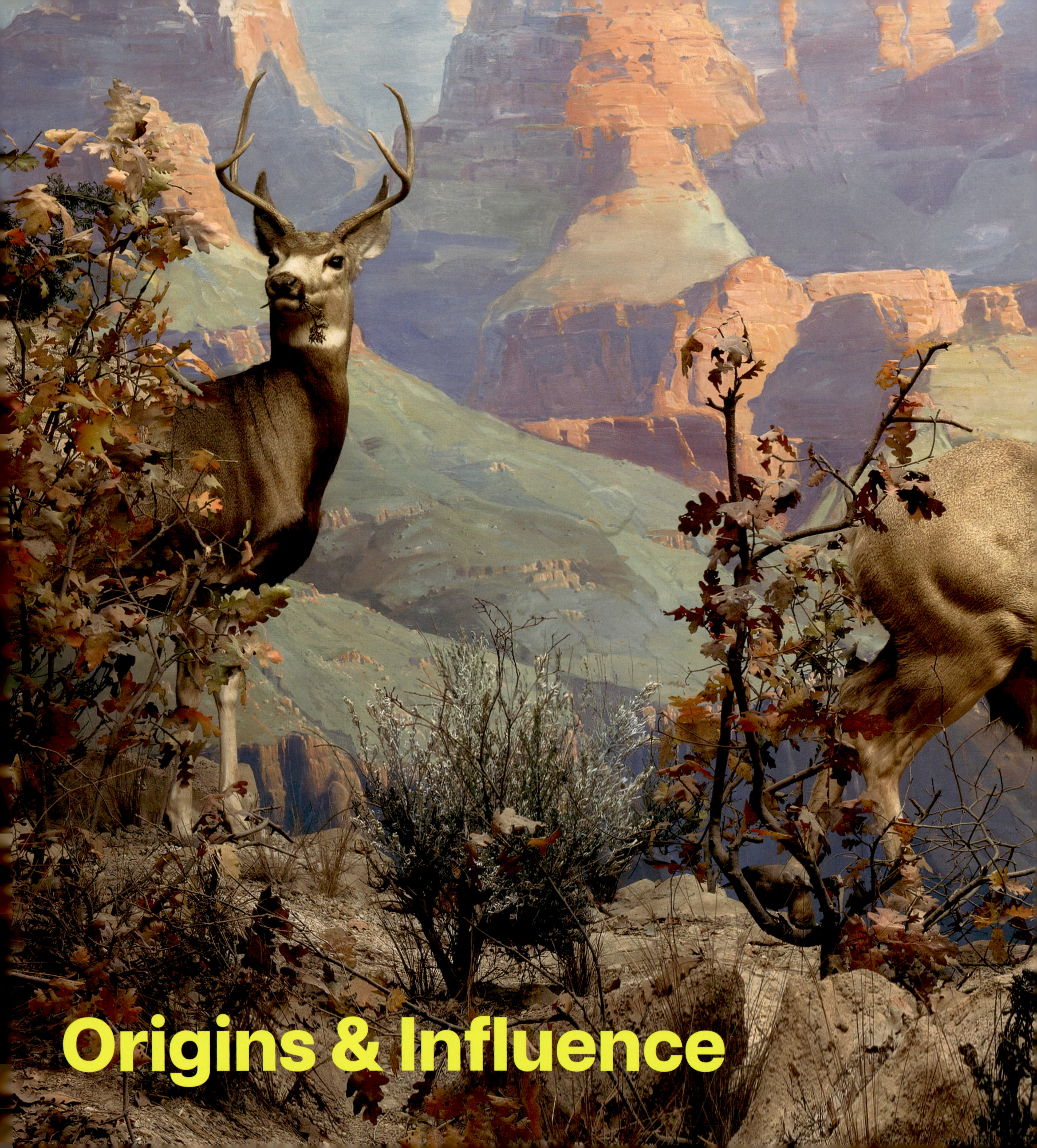

Origins & Influence

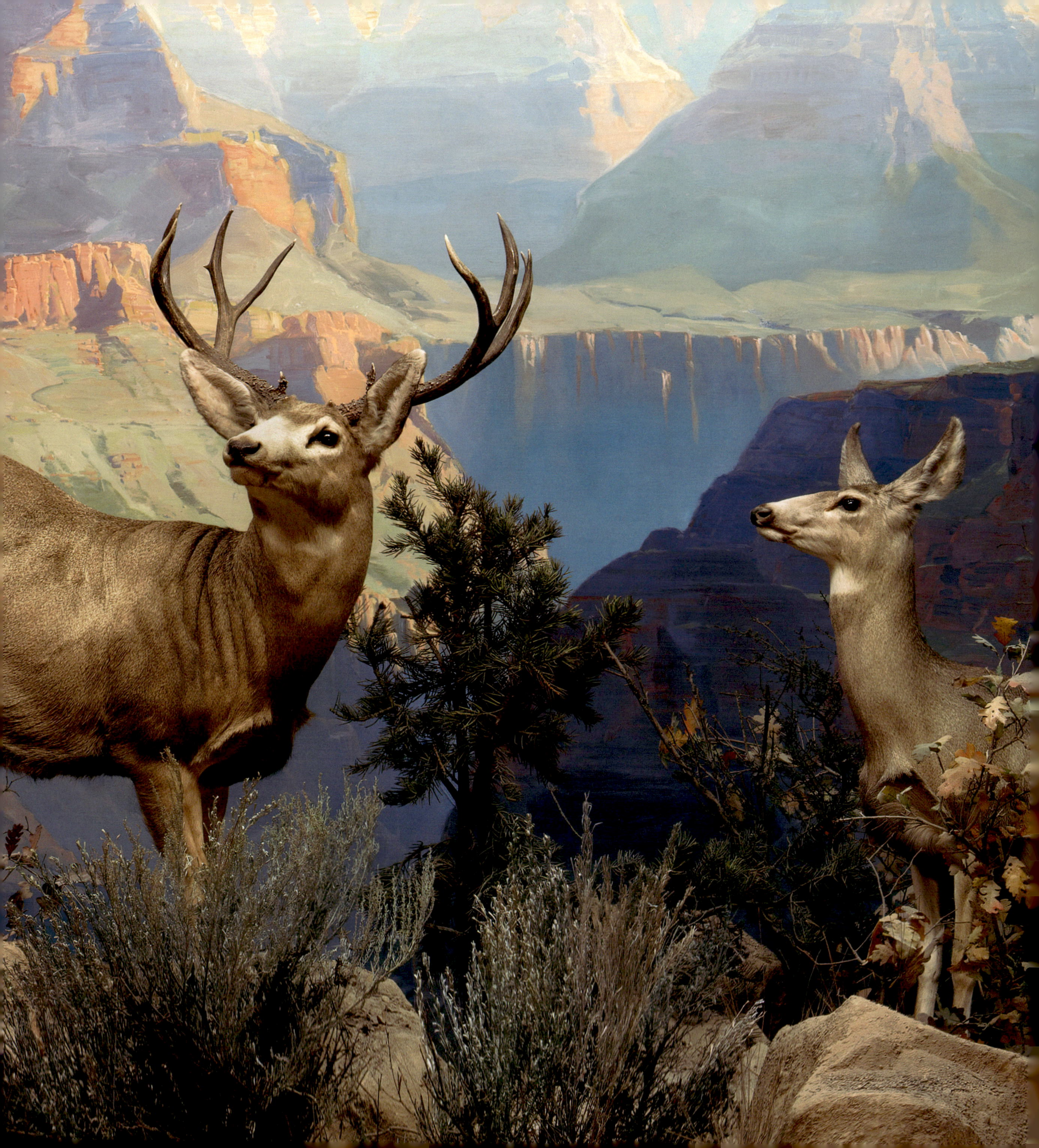

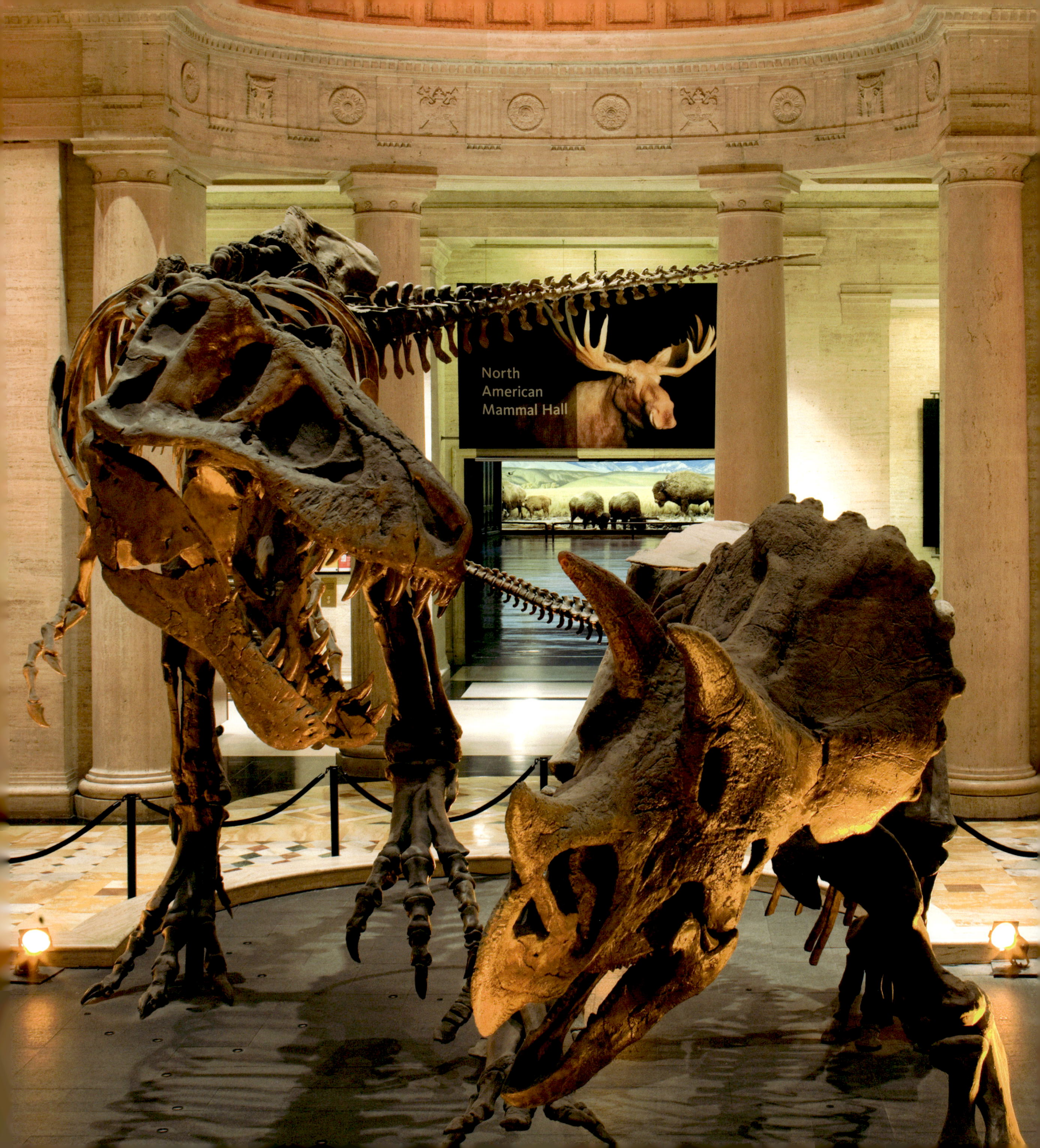
North
American
Mammal Hall

Introducing Dioramas

Matt Davis

Think of a natural history museum and you'll probably think of two things: dinosaurs and dioramas. Known as habitat groups, these dioramas show lifelike animals in re-creations of their natural environments using three essential components: a realistic background painted on a curved wall, a modeled foreground of leaves and rocks, and life-size taxidermied animals. A habitat group mixes together art and science into an immersive scene, a realistic illusion of a single moment frozen forever in time. As Senior Curator of Habitats Melville Lincoln wrote in 1941, "Science furnishes the material, art the finished picture."[3]

Although dioramas may seem synonymous with natural history museums, that was not always the case. When the Natural History Museum of Los Angeles County (NHM) first opened to the public in 1913, it didn't even have any dioramas. The traditional, scientifically accepted way of displaying specimens was to line them up in neat rows of taxonomic rank (kingdom, phylum, class, order, family, genus, and species) in tightly packed glass cases so that visitors could see the physical differences and similarities between many species. Birds were with birds (class: Aves), mammals with mammals (class: Mammalia), skulls in one display, shoulder blades in another. The skeleton of a pinyon jay (a bird), the antlers of a mule deer (a mammal), and the dried cone of a pinyon pine (a plant)

Previous spread
Opened in the late 1920s, the *Mule Deer* group was one of the first dioramas at NHM. Its spectacular setting along the North Rim of the Grand Canyon exemplified how revolutionary the dramatic, lifelike scenes of dioramas were compared to traditional display methods (2014).

← A view from NHM's foyer captures both dinosaurs and dioramas. Unlike the skeletons of prehistoric animals, habitat dioramas were actually a later addition to natural history museums. They first became popular during environmental crises in the early 1900s (2010).

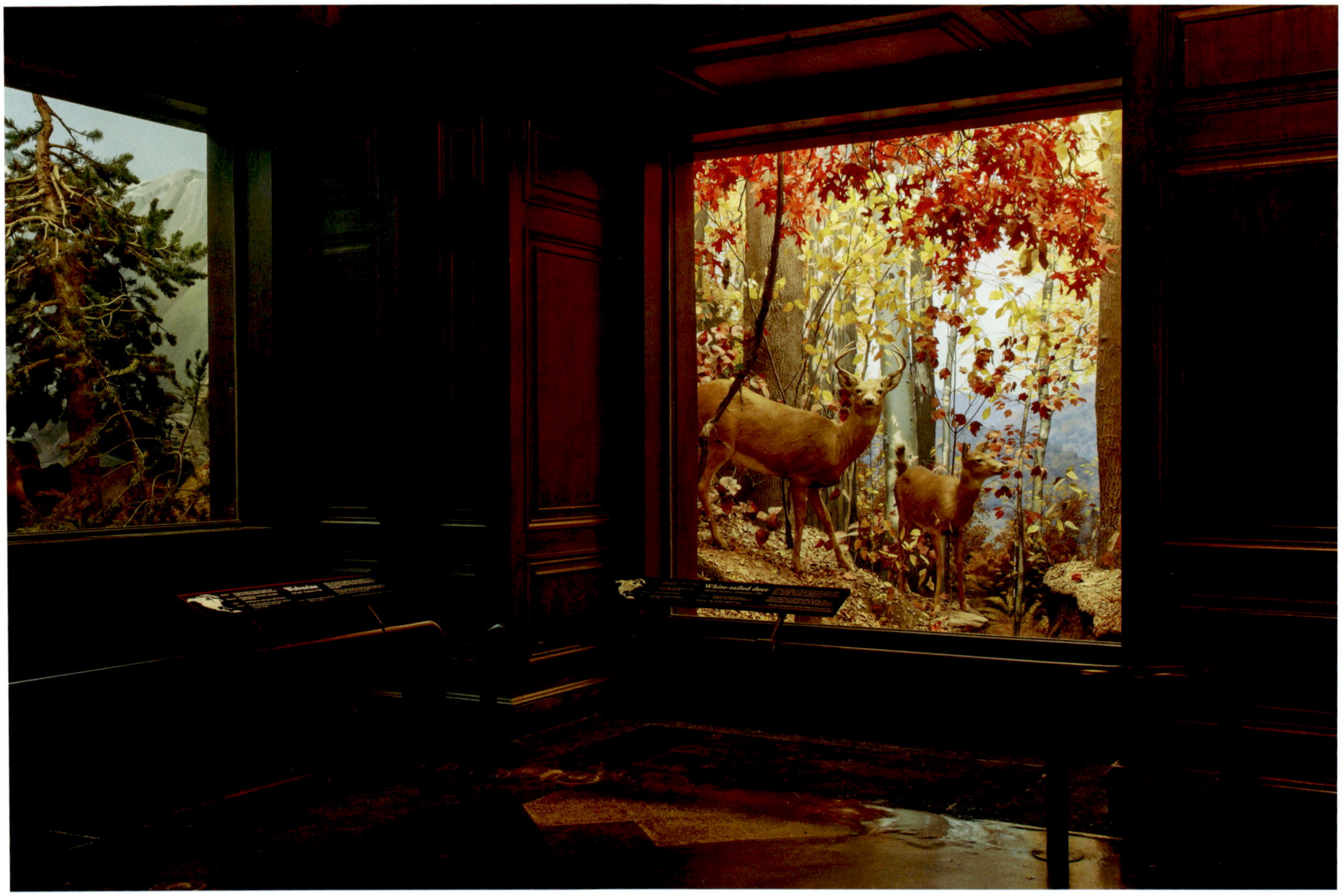

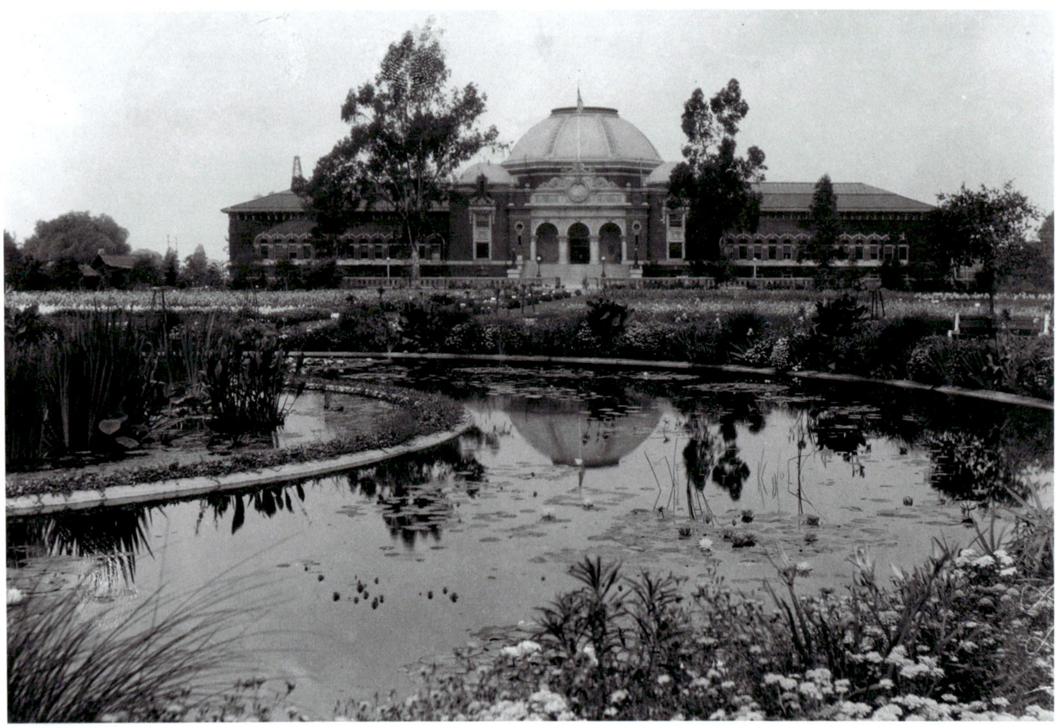

↑ This scene of white-tailed deer stepping warily through an autumnal Pennsylvania forest is an excellent example of the habitat group diorama. Curved painted backgrounds, modeled foregrounds, and taxidermy combine to reveal habitats frozen in time (2016).

← NHM opened in 1913 in Exposition Park, before dioramas exploded in popularity.

would all have been exhibited far apart from each other in different cases because they weren't related, even though the three species live together in the same habitat along the rim of the Grand Canyon.

In the early 1900s, taxidermists and educators began to question whether traditional exhibition cases full of lifeless bones were really the best way to educate and interest the public. A new field of science called ecology, which investigates the connections between species and their habitats, was developing. Rows of bones were good at showing differences in anatomy, but they couldn't show museum visitors the ecological relationships between species that lived together in the same habitat, like pinyon jays, mule deer, and pinyon pine trees. The deer used the trees for shelter, while the jays built nests in the trees and fed on their seeds. In exchange, the trees' seeds were dispersed far and wide by the jays. By showing animals living together in their natural habitats, dioramas could be a revolutionary and poetic

↓ Habitat groups were a major revolution contrasting with traditional exhibitions like the museum's original Hancock Hall of Science, which displayed bones in neat taxonomic rows (1925).

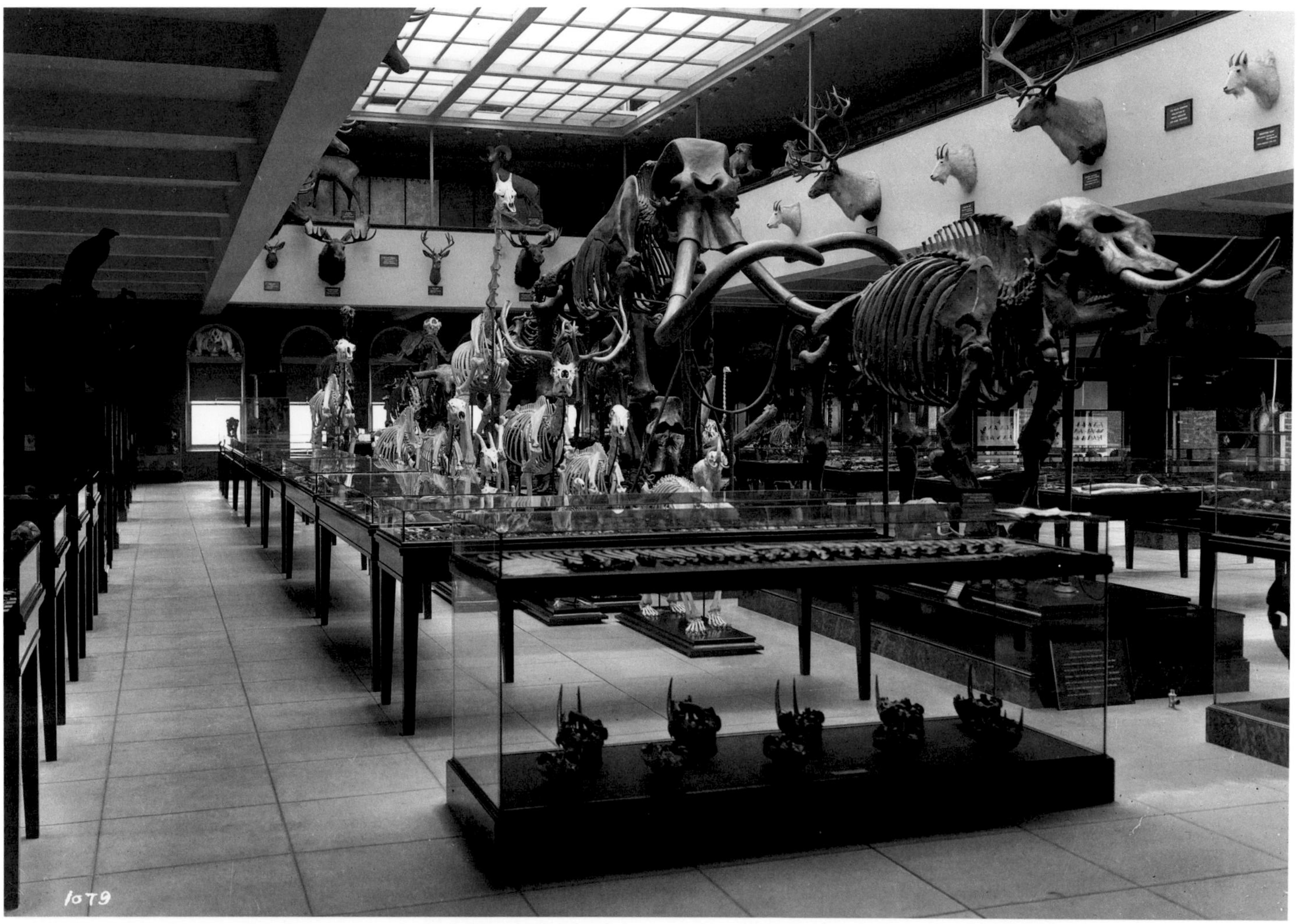

acknowledgment that everything—from wind and water to rocks, trees, and birds—is connected.

The environmental catastrophe cascading across North America during the early 20th century was even more important for the development of dioramas. Wanton hunting and habitat destruction were driving many iconic species like bison extinct. By 1895 the vast herds of tule elk that once roamed California had dwindled to just 28 individuals. Mule deer were predicted to disappear before 1915. By 1925 Californians had exterminated the California grizzly bear—the animal emblazoned on their own flag. The situation was even worse for birds. The contemporary fashion for voluminous feathered hats led to "plume hunters" slaughtering wild birds by the thousands to secure feathers for markets in Europe and North America. Passenger pigeons once had flocks so large they blocked out the sun. The last of their kind died in the Cincinnati Zoo in 1914. Four years later, in the same cage at the same zoo, the last Carolina parakeet would follow.

This rapidly growing list of extinctions weighed heavily on the biologists and artists who would soon create NHM's diorama halls, such as Museum Director William Bryan and diorama Background Artist Charles Abel Corwin. In 1911, before they worked at NHM, Bryan and Corwin (along with taxidermist Homer Dill of the University of Iowa) went on an expedition to the tiny Hawaiian island of Laysan to resurvey its unique biodiversity, which Bryan had studied eight years earlier. Where before Bryan had seen an immense breeding colony of seabirds, he now found a hellscape. Right before his arrival, poachers had slaughtered over 300,000 Laysan albatrosses for their feathers, leaving the tiny island covered in bones and rotting carcasses. The plume hunters stuffed hundreds of living birds at a time into a cistern, leaving them to starve to death so that their skin and feathers would be less greasy from fat deposits and thus easier to clean. Bryan was shocked. "Not a single bird remains, while heaps of the slain lie as mute testimony of the awful slaughter."[4]

What were museums doing to save wildlife? How could they get an increasingly urbanized public to care about species in the rapidly changing American West, let alone on a tiny, isolated island like Laysan? America's national and state park systems were still in their infancy, and there were few examples of successful conservation programs. The prevailing sentiment was that we had already passed the point of no return and that many

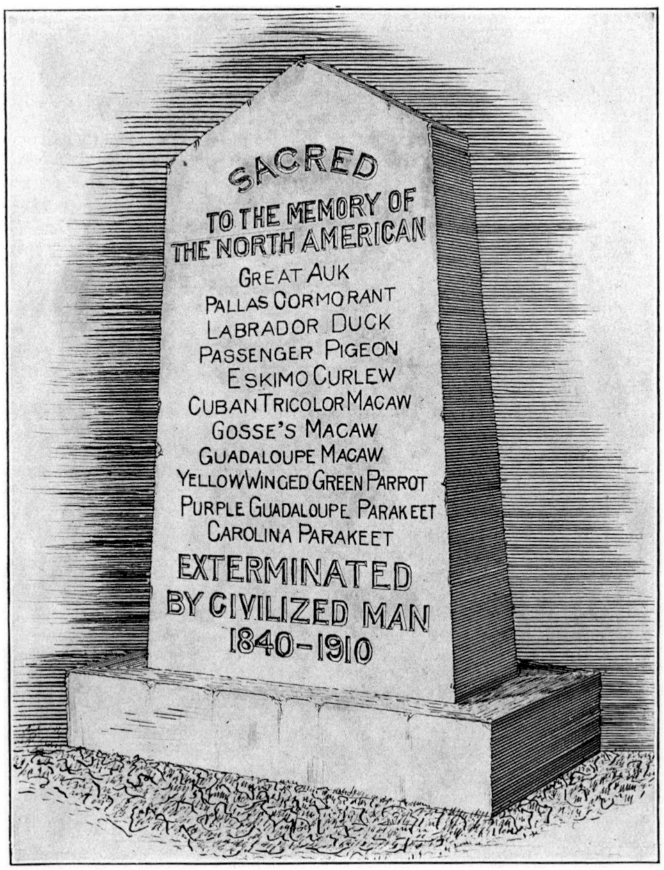

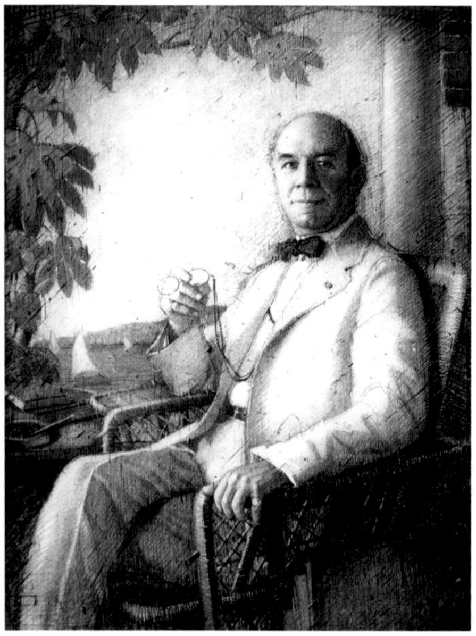

← A tombstone illustration from conservationist William T. Hornaday's influential 1913 book *Our Vanishing Wildlife* laments the rapidly growing number of bird extinctions.

↑ William Bryan was horrified by the industrial scale of wildlife slaughter he saw as a biologist. He went on to oversee the creation of four large diorama halls at NHM during the 1920s as the second museum director (1927).

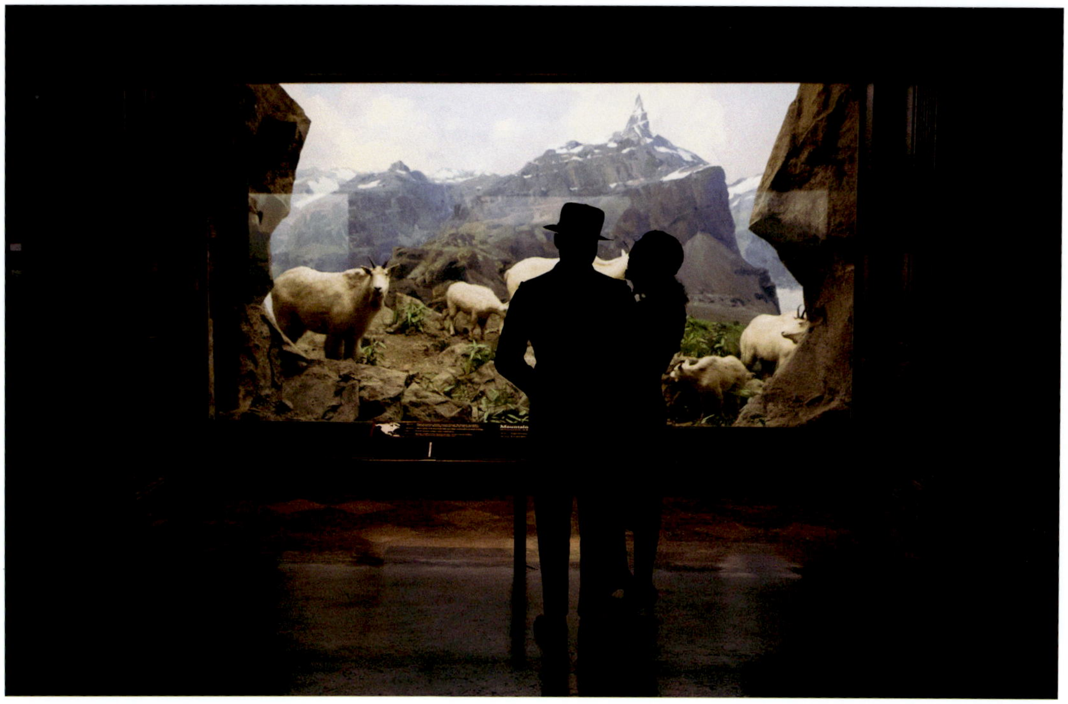

→ The *Mountain Goat* group, shown with historical reenactors, looks almost the same as at its opening in the late 1920s. The realism and theatricality of dioramas amazed early visitors and helped transform American museums into places that welcomed everyone, not just well-educated specialists (2018).

species would soon disappear forever under the banner of progress. Faced with overwhelming environmental destruction, artists rushed to preserve entire habitats behind glass in museums, creating a new, fatalistic art form: the habitat group diorama. It was an art form that NHM wholeheartedly embraced in the 1920s, building major expansions to reorient the museum around four new diorama halls. The designer of these halls, pioneering Taxidermist John Rowley, explained their purpose to Angelenos: "Man is the exterminator, but before it is too late it is the intention of those who are responsible for it that the [diorama halls] shall typify and leave permanent records of these remarkable scenes."[5] If the new dioramas did not inspire visitors to conserve wildlife, then at least they would preserve these extinct landscapes in the public memory forever, some small simulacrum of the Eden that society had so blindly forsaken.

But visitors *were* inspired, amazed even. Museums reported more visitors and more donors with each new diorama they opened. At the University of Iowa Natural History Museum, Dill and Corwin built a cyclorama (a circular diorama) out of the specimens collected on their trip to Laysan with Bryan. When the *Laysan Island Cyclorama* opened in 1914, the museum was mobbed by visitors; one out of every five people in Iowa City went to see this immersive re-creation of the tiny, endangered Hawaiian island. "Who are these people?" Dill rhetorically asked museum leaders at a national conference in 1915.

> Some of them are students, some are farmers, others are members of nature clubs from neighboring towns visiting the place in a body, and still others are mothers with their children. Here are all kinds of people! Students, men of science, and people who would never read a book or go to a lecture are here learning something.[6]

Dioramas were revolutionary explosions of color and drama compared to the dull exhibits of lifeless bones visitors were familiar with. Even today, with all our modern technology, dioramas still inspire wonder and pull us in with their convincing realism. Imagine how exhilarating it would have been to face a leopard, frozen mid-snarl, in a time before computers, TV, or even inexpensive color books. When they opened, the diorama halls at NHM were probably the only way most Angelenos would have ever seen animals like mountain goats or sitatunga antelopes. The 1891 prophecy of early taxidermist and conservationist William T. Hornaday had finally come true. "To-day the average museum is as lifeless as a dictionary," he wrote, but with dioramas, ". . . the museum of the future will be life itself."[7]

Diorama halls spread quickly across the country as museums raced around the world to collect specimens before their habitats disappeared. Everything in the environment, from large mammals to rocks and bugs, needed to be collected to reproduce a convincing scene.

The specimens that didn't make it into the dioramas began to form the foundation of many museums' now invaluable scientific research collections. Dioramas were also a significant part of the early conservation movement; the taxidermists like Hornaday and Rowley who searched out animals for dioramas were keenly aware of how rare they were becoming. After multiple expeditions up and down the California coast in the early 1900s, Rowley alerted scientists to the dire situation facing marine mammals and lobbied successfully for a state law to protect seals and sea lions on Santa Barbara Island. Dioramas, like Rowley's *Steller Sea Lion* group, grabbed the public's attention around endangered animals and landscapes, as if the mere act of placing a habitat inside a large glass box made people realize it needed to be saved. Habitat groups at museums around the country led directly to a national program to protect bison, America's first federal bird reserve, and Africa's first national park.

The conservation movement worked; there are many more bison, tule elk, and mule deer alive now than when these animals were first placed into NHM dioramas 100 years ago. But while dioramas did inspire visitors to protect wildlife, they also embodied a particular view of wilderness that conservationists and communities are still struggling with today. NHM's *Mule Deer* diorama opened in the late 1920s to introduce Angelenos to the pristine and majestic wilderness of the country's new Grand Canyon National Park. Yet, like most scenes in the dioramas, it was a "wilderness" that was created, not found. Conservation laws and parks favored rich sportsmen, tourists, and spectacular views, while brushing aside the poor settlers and Indigenous peoples that inconveniently already lived in these "pristine" habitats. When the winter homes of the Havasupai people were suddenly placed within Grand Canyon National Park by a line on a map, traditional activities like hunting mule deer and gathering pinyon pine nuts for food became "poaching" overnight, punishable by a year in prison. These laws didn't stop a wealthy sportsman in 1924 from getting a permit and a tax write-off to hunt mule deer for the museum's diorama, though, nor did they even save the mule deer themselves. Freed from their traditional predators by the government's heavy-handed conservation program, mule deer exploded in population and then starved to death once they had devoured all the food in the area. Placing habitats inside dioramas galvanized the public to protect wilderness, but it also made them think that nature was something they could only look at and never touch.

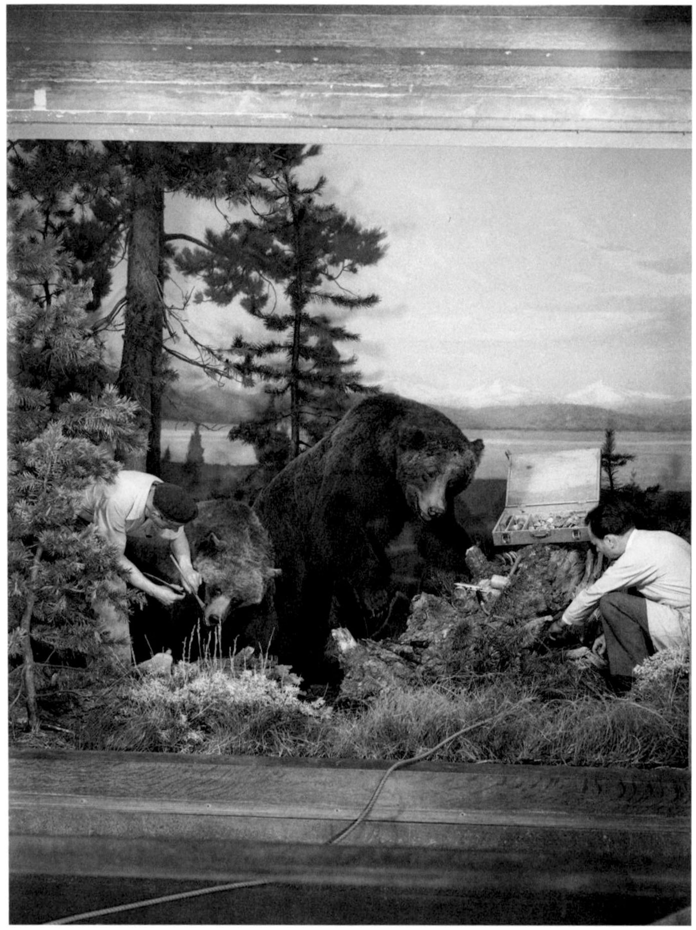

↑ Unlike those of other museums, NHM's dioramas have large sash windows that have allowed frequent updating during their 100-year history to make them more attractive and scientifically accurate (1945).

Once revolutionary, habitat dioramas quickly became standard and, eventually, old-fashioned. Museums began to wonder if dioramas were really worth the time and incredible expense they took to build. Did visitors even understand all the subtle ecological stories hidden in these pleasing scenes? By design, they never had long explanatory labels—much was left up to the visitor's own exploration and interpretation. And how did visitors interpret dioramas—as faithful records of wilderness before humans despoiled it? They weren't exact duplicates of nature; they were works of art, created first in the minds of artists. They showed an imagined world without humans, though our footprints are hidden everywhere in the dioramas if you know where to look. The tule elk rest not in the vast tule marshes that gave them their name but in the parched scrublands that sprung up after these marshes were drained for agriculture.

Faced with these questions and increasingly negative views toward hunting big game, some museums abandoned their dioramas, relegating them to the realm of dusty historical curiosity or dismantling their halls altogether. But there is one thing that separates the dioramas of NHM from those of all other museums: they open up. Unlike other museums, where dioramas were made once, decades ago, and then sealed behind glass forever, NHM's dioramas are covered with large sash windows that can be easily opened and closed. NHM staff change the habitat groups frequently to improve their scientific accuracy and artistic value. They weren't all made 100 years ago; they were built and rebuilt over the last century, each new diorama tracking changes in technique and public sentiment. Old taxidermy mounts from big game hunters are swapped out, new animals donated by zoos are put in, backgrounds are repainted, habitats are made closer to what they actually are, not what we wish them to be.

Rather than abandon dioramas, NHM has continued to embrace them, keeping the art form alive, creating habitat groups, training the next cohort of taxidermists, and enchanting each new generation of Angelenos. The halls aren't just habitats for wildlife anymore; they have become places for humans too. The large, open spaces of the habitat halls have been used for everything from bomb shelters and community organizing to wedding venues and trendy dance clubs. The dioramas still have a special pull on all of us. While every other museum exhibition has come and gone, these halls alone have remained standing for over 100 years. This is their story.

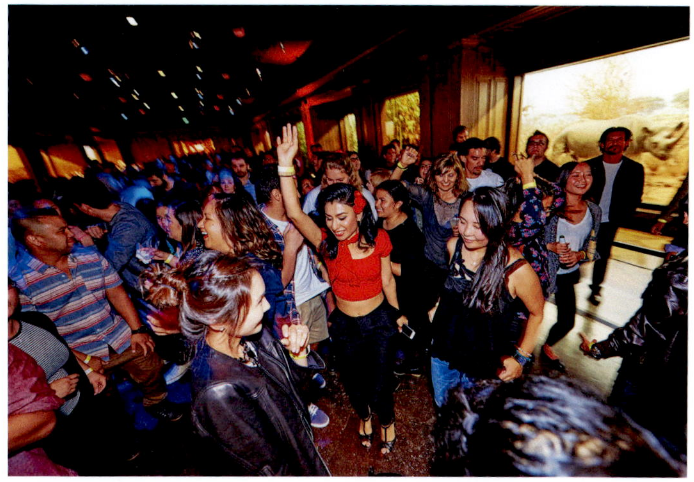

↑ As the largest open spaces in the museum, the diorama halls frequently host community gatherings such as NHM's popular First Fridays program (2015).

↓ Sparkle Juarez demonstrates taxidermy to museum visitors, showing how NHM continues to build new habitat groups and train the next generation of diorama artists (2019).

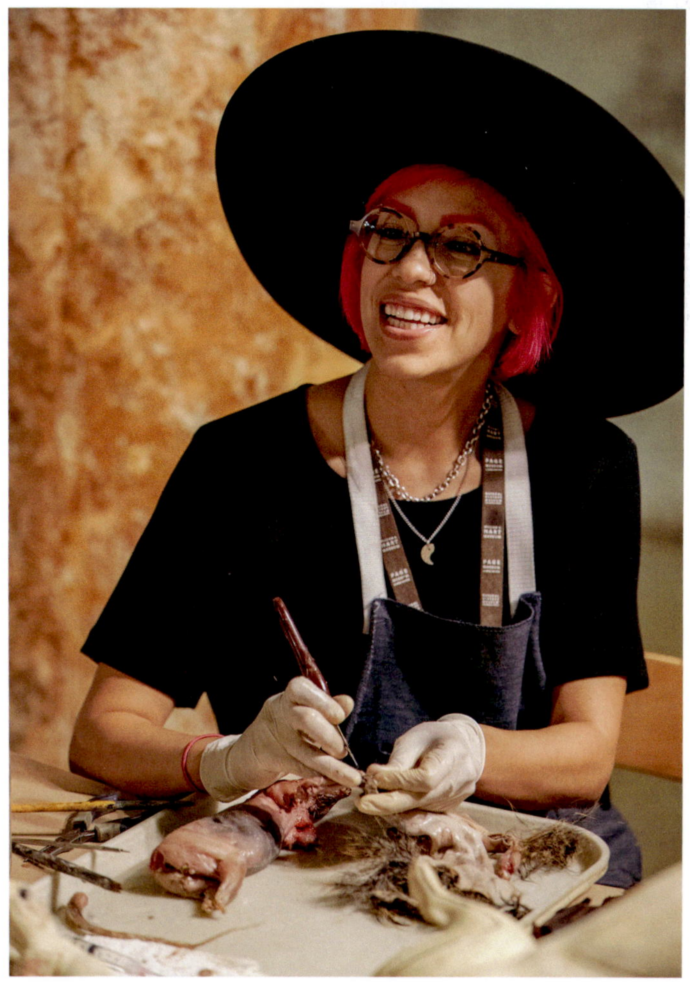

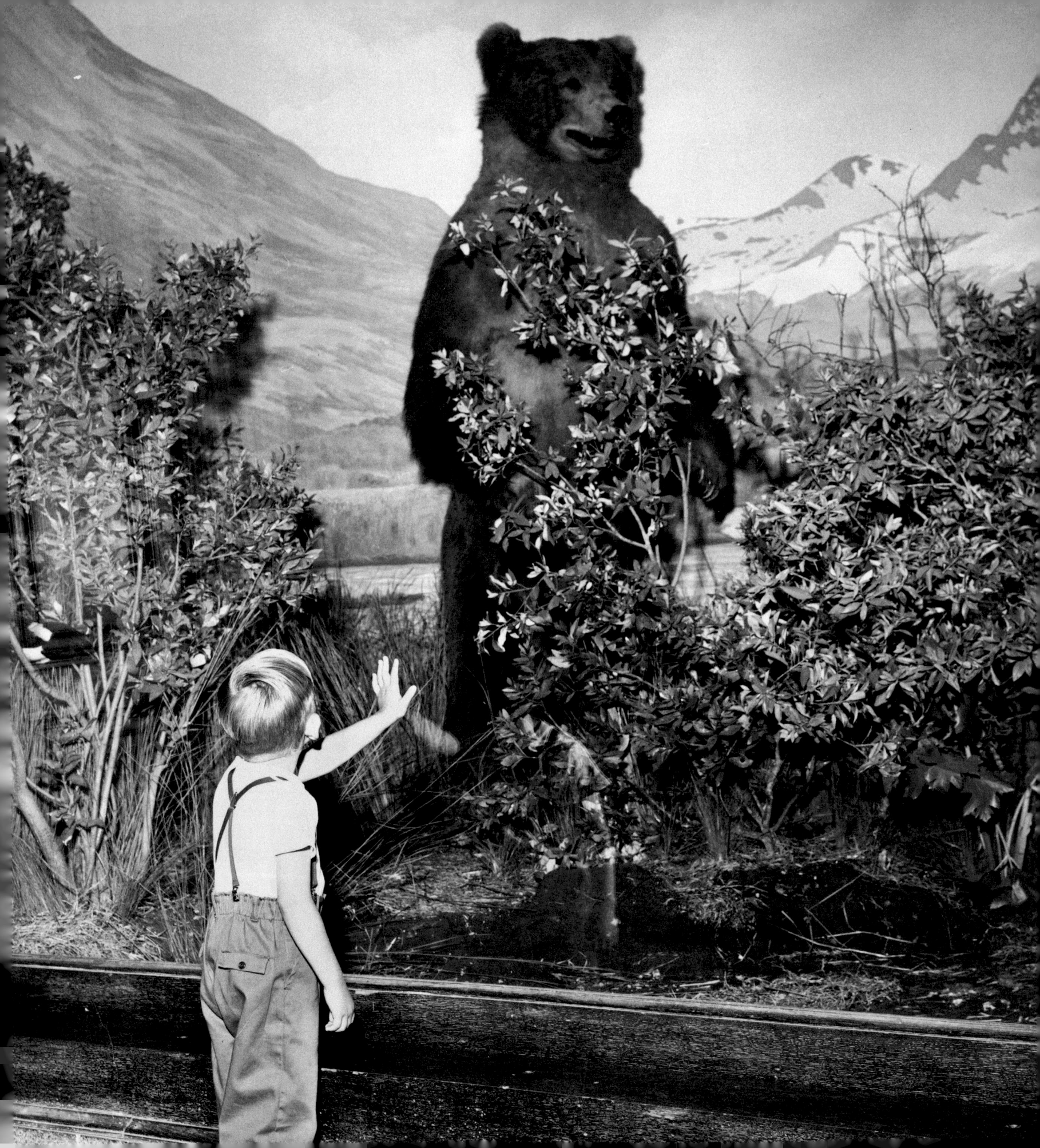

Thinking with Dioramas

Jon Christensen and Ursula K. Heise

← A child presses against the glass of the museum's first *Kodiak Bear* group (1950s).

As the famous rock singer Sheryl Crow snarls her way through "If It Makes You Happy," the official music video for her most popular song shows her trapped inside an NHM diorama of a simple red room wedged between a taxidermied wolf pack in the mountains on one side and polar bears on ice floes on the other.

As the camera moves around the diorama hall—an amalgam of different dioramas from across the museum is used to create a fictional "Endangered Species" hall—we see grizzly bears, moose, elephants, walruses, caribou, rhinoceroses, foxes, and other animals from around the world behind glass. In the video, streamed more than 59 million times on YouTube and still popular decades after the song won a Grammy in 1997, the diorama hall functions as a metaphor of enclosure, as well as transgression, escape, and freedom. Exhibit Closed and No Photography signs control access to this kingdom of dead animals, along with a sleepy museum guard, even as the camera leads us into the exhibition anyway. A visitor shoots a photo, and a very much alive elephant lumbers through the hall with an odd assortment of other visitors. Crow wears animal-print clothes and claws comically at the camera. Fake trees and animals decorate her prisonlike room. A small window in the back wall seems to look out on a perhaps real forest.

When visiting Boy Scouts press their faces against the diorama glass as if they were trapped inside, wanting out, and a girl wears braids that look like the horns of the Arabian oryx that she is contemplating, we may begin to wonder which side of the glass is holding back the wildness of nature.

The music video is a testament to the ongoing popularity of dioramas as well as to the different creative meanings that dioramas can have over time. In the early days, diorama creators were concerned that wild nature was disappearing from the world. They wanted to preserve examples of wild nature to inform, educate, and improve mankind, even if it meant they had to kill animals and put them behind glass to do so. So hunting parties went out and shot animals and shipped them back to the museum, where taxidermists resurrected them to appear alive in their natural habitats. Those habitats were often constructed from a combination of plant specimens collected in the field where the animals died, along with other specimens from nearby that were close enough in appearance to create the right effect. Painted backdrops completed the picture, while usually erasing any trace of human beings in the landscape.

This cultural history still resonates today even as dioramas are appropriated for new contexts and meanings. The boundary between reality and art is always uncertain in dioramas, and often in how they are represented, too. In the 1970s, photographer Hiroshi Sugimoto became famous for his Dioramas series. These photographs of habitat groups, including some at NHM, leave out any

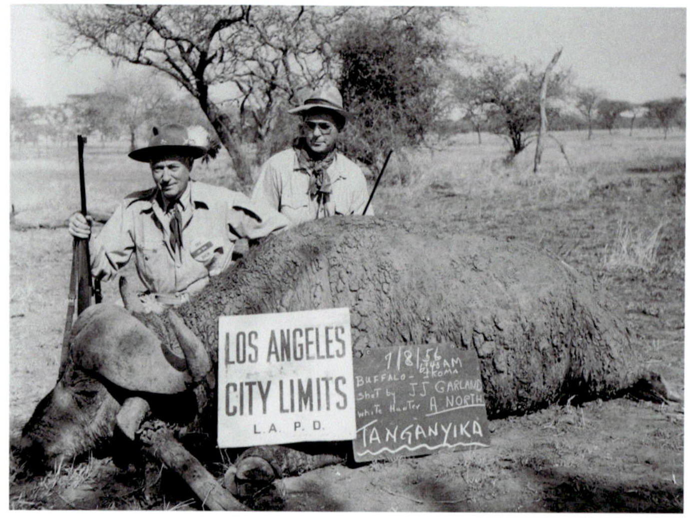

↗ The legacy of the "great white hunter" and colonialism hangs heavy over dioramas. "White hunter" (listed on the sign in this image), a typical role on safaris along with cook or driver, designated the lead hunting guide of European descent. Native Africans also served as gun bearers, scouts, and hunters on safaris, but they were never afforded the high status and colorful, adventurous identity given to the white hunter. Museum benefactor John J. Garland (*left*) and Dr. Henry Morgan (*right*) shot this Cape buffalo under the guidance of white hunter Alan North on an East African expedition to gather specimens for the dioramas (1956).

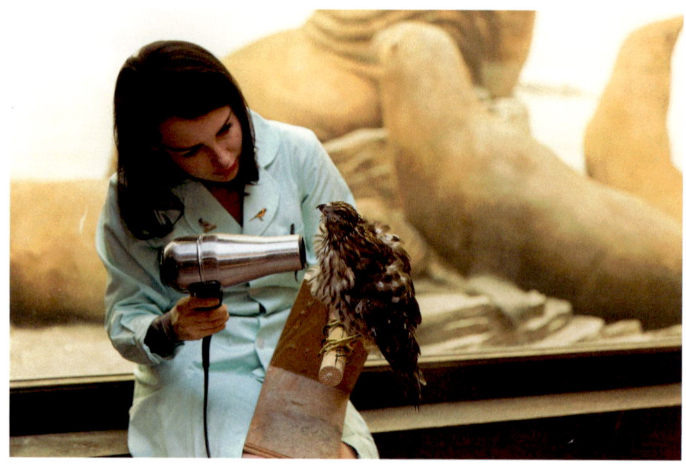

→ Diorama specimens need their own stylists, such as Taxidermist Allis Markham, to get ready for their close-ups (2018).

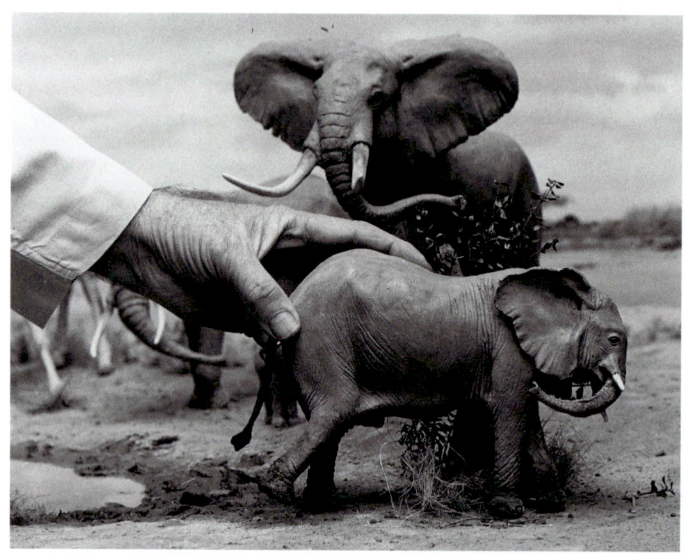

→ The shaping hand of man is clear in this photograph of a taxidermist arranging the maquette for the future *Watering Hole* diorama (1965).

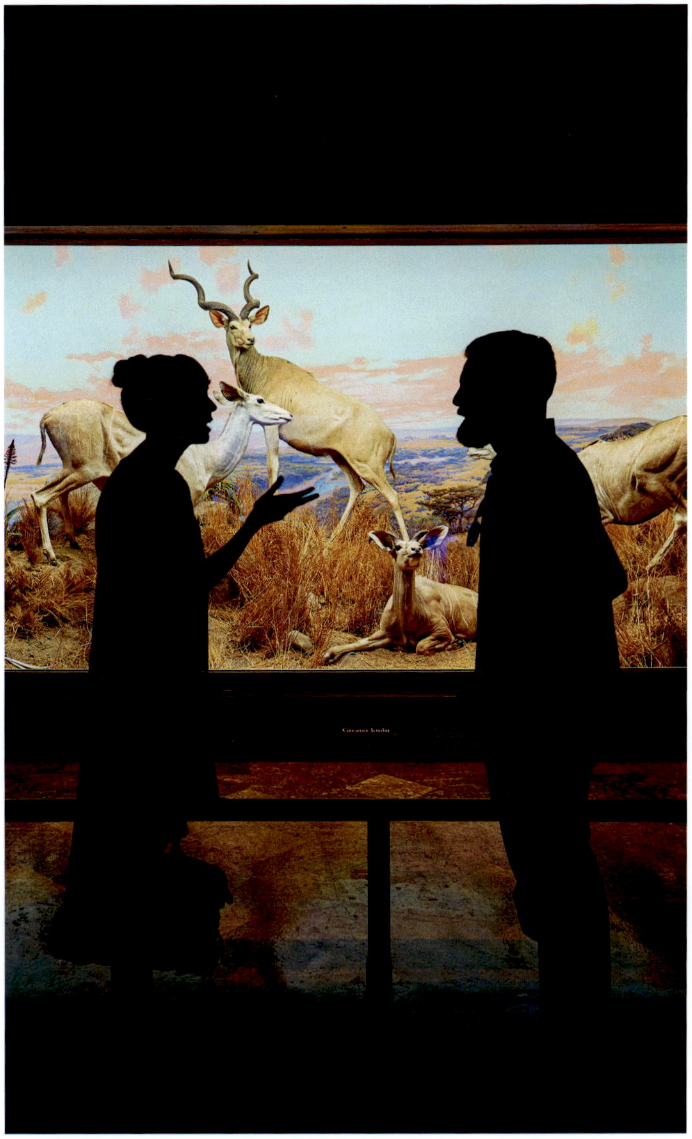

↑ These greater kudu have ignited a date night conversation at a First Friday event, one that they seem eager to eavesdrop on as well (2015).

evidence of enclosure or signage. At first glance they look like genuine wildlife photography. In a 2012 Super Bowl commercial seen by 111 million viewers, Matthew Broderick parodied his own role as Ferris Bueller. Playing hooky from filming a commercial, he visits the *Walrus* diorama at NHM. Broderick contemplates the walrus while the walrus watches him, evoking a scene in *Ferris Bueller's Day Off* and one of Bueller's most-quoted pronouncements:

> Isms, in my opinion, are not good. A person should not believe in an ism, he should believe in himself. I quote John Lennon: "I don't believe in 'Beatles,' I just believe in me." Good point there. After all, he was the Walrus. I could be the Walrus. I'd still have to bum rides off of people.[8]

In Broderick's Super Bowl encounter with a taxidermied walrus in a diorama, reality, fiction, and fiction-upon-fiction can no longer be easily distinguished.

Natural species are "good to think with," the French anthropologist Claude Lévi-Strauss famously wrote in his book on totemism.[9] And animals in dioramas continue to unsettle our thinking with their uncanny mix of reality and artifice. The animals and many of the plants in dioramas are real, but they are transformed through a mash-up of science, art, and culture—a combination of biology, ecology, taxidermy, sculpture, painting, composition, writing, history, and changing cultural ideas and narratives. The material product of this mix of disciplines highlights how ideas that originate in one field reshape ideas in other fields.

Representations of nature in diorama backdrops have been influenced by traditions of landscape painting and movie backdrops, with their conventions of visual realism, compositional balance, and creation of particular moods through dark woods, looming mountains, sunny skies, or threatening clouds. Interpretive labels have sought to shape what people notice in dioramas and think about as they move from one to another. Changing interpretations have, in turn, shaped what gets included in dioramas. The miniature ecosystems they represent shaped ideas about what makes up real ecosystems, as if they too could be bounded by a container.

For attentive visitors today, this fascinating mix of reality, representation, culture, knowledge, and aesthetics raises endless questions. In a 2012 episode of *Mad Men*, a girl visiting New York City's American Museum of Natural History—filmed, in another twist on reality and representation, at NHM—asks about the *Muskox* diorama, "How'd

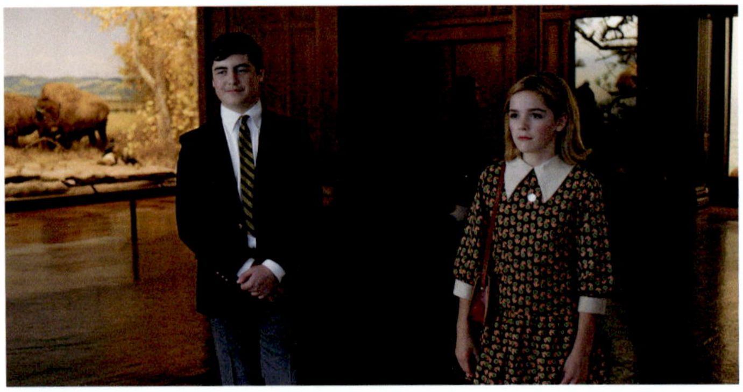

← Sally Draper (*right*) discusses dioramas with Glen Bishop (*left*) in a scene from the TV show *Mad Men*. Although set at the American Museum of Natural History in New York City, like many shows and movies featuring dioramas it was actually filmed at NHM.

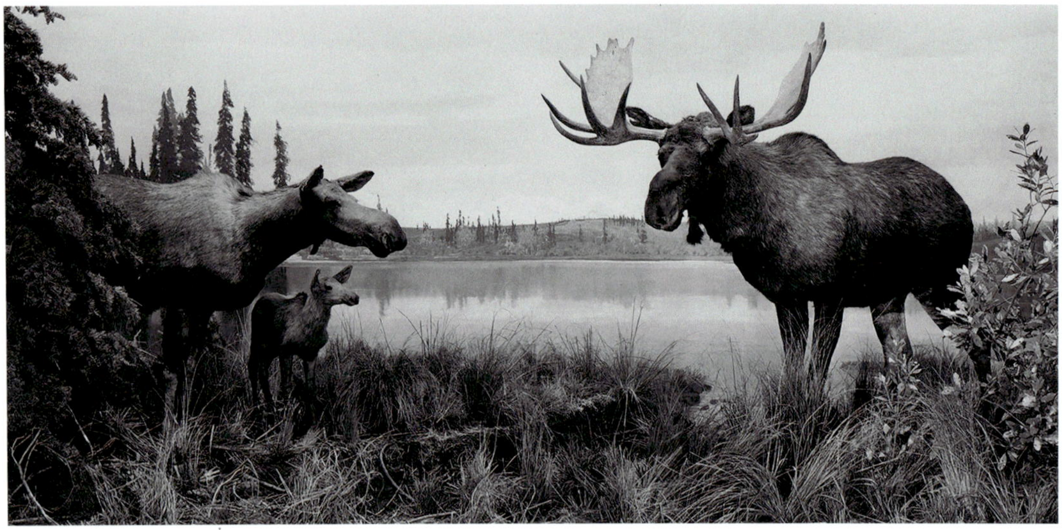

← It is unlikely that a cow moose would tolerate a bull moose so close to her young calf, but the diorama designers needed "Dad" to complete a family portrait modeled on a human ideal (1950s).

they get all these animals?" Her friend dryly responds that "Teddy Roosevelt killed them." "Are they a family?" the girl wonders, to which her friend quips, "I hope so. Otherwise, what were they doing? Walking around saying, 'We just need a baby to finish this diorama'?"[10]

The exchange is a concise, humorous take on cultural studies scholar Donna Haraway's feminist critique of dioramas in her essay entitled "Teddy Bear Patriarchy: Taxidermy in the Garden of Eden." Animal groups have often been staged as nuclear families in dioramas, with an alpha dad, a submissive mom, and a playful brood of children. Today, forcing radically different species to conform to our own ideals of familial harmony has become problematic, as the girl in *Mad Men* suggests. We know there are many different types of families among species, whether human or not. At the same time, these ideal animal families protected behind glass retain the nostalgic charm of what some call a "heterotopia"—a small utopian bubble in the midst of the fractured, contentious, and exploitative social relationships that *Mad Men* so eloquently portrays.

Dioramas are themselves historical artifacts. Some of this history may not be obvious to the average visitor and might sometimes be uncomfortable to engage with, especially their connections with racist, colonial histories. While the early makers of dioramas were concerned about preserving pure representations of wild nature in museums before they disappeared in the wild, many, like the Smithsonian's influential taxidermist William T. Hornaday, were also connected to ideologies in science and society that were concerned with preserving a "pure" white race of human beings. They were worried that the white race might go extinct too, along with the passenger pigeon, either through "mixing" with other races or through dissipation, too much luxury and laziness. Promoting manly activities such as hunting was seen as a way to stave off "race suicide," in addition to providing specimens for dioramas.

Sometimes, difficult but important conversations can be sparked by simply changing the wall labels that accompany dioramas. In 2018, for example, the American Museum of Natural History in New York City modified a

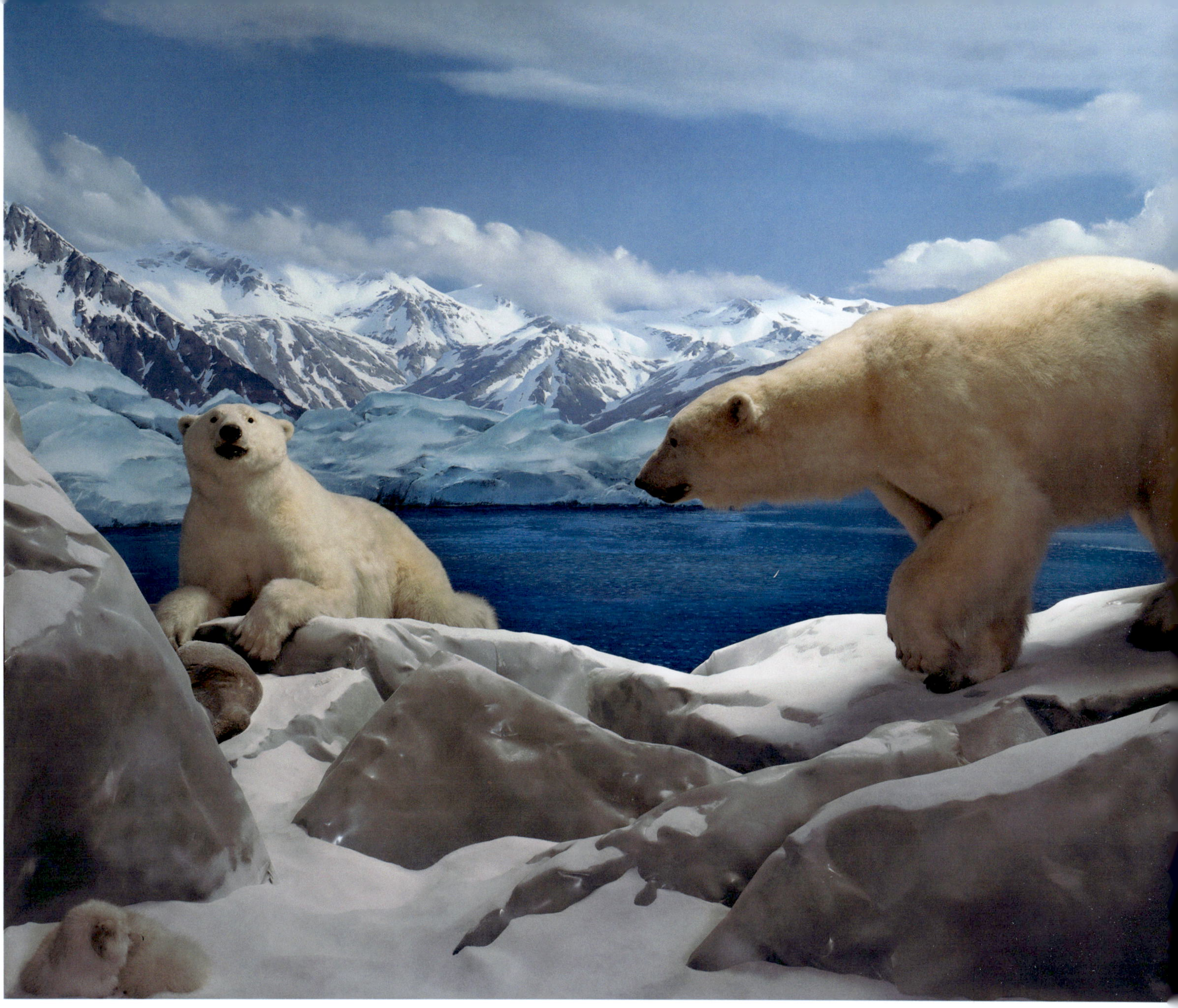

↑ These polar bears carry a very different and weightier meaning today than when they were first placed in this diorama in 1976 (2014).

diorama depicting an imaginary meeting between 17th-century Dutch settlers and a group of Lenape Indigenous people by adding overlays to the glass that explained the ways the scene was historically inaccurate. These overlays both preserved the integrity of the original diorama as a historical artifact and opened up conversations about its political and racial history.

Sometimes dioramas can change meaning without any overt intervention. A traditional diorama of polar bears on an ice floe hardly needs any changes in presentation or interpretation to take on new meanings in our era of climate change. NHM's now seems to have eerily anticipated global warming.

New diorama designs can also open up new conversations. A diorama showing native and nonnative urban wildlife in a Los Angeles backyard—a coyote with a cat in its mouth, birds around a feeder, a rat scurrying away, the downtown skyline in the hazy distance—illustrates the relationship between human residents, domesticated animals, and wild animals, all living together in close proximity. This is what ecologists call a "novel ecosystem," a habitat shaped by both people and nature.

But these examples also hint at the limitations of dioramas as a form. By that we mean not the content of the diorama but the formal structure that contains the content—an art form like a painting, a sculpture, or a movie. Formally, a diorama is one frozen scene in a narrative or story, like one still frame in a movie or one panel in a graphic novel. Often that scene evokes a well-known story or what literary scholars call a masterplot. We call these masterplots because they often dominate the ways people think about things, sometimes in very unthinking ways, as we've seen in representations of animal families in dioramas. A single scene of a pride of lions can stand in for a whole larger narrative of the hunt, the hierarchy, the lion king. But while it is easy to call up a masterplot with just one scene, it is much harder to disrupt or question that dominant narrative with just one scene. This sticky effect can overpower labels, other displays, and programming, which may just be ignored as visitors immerse themselves in a diorama scene that is by its nature very compelling, filled as it is with life, even in a preserved form.

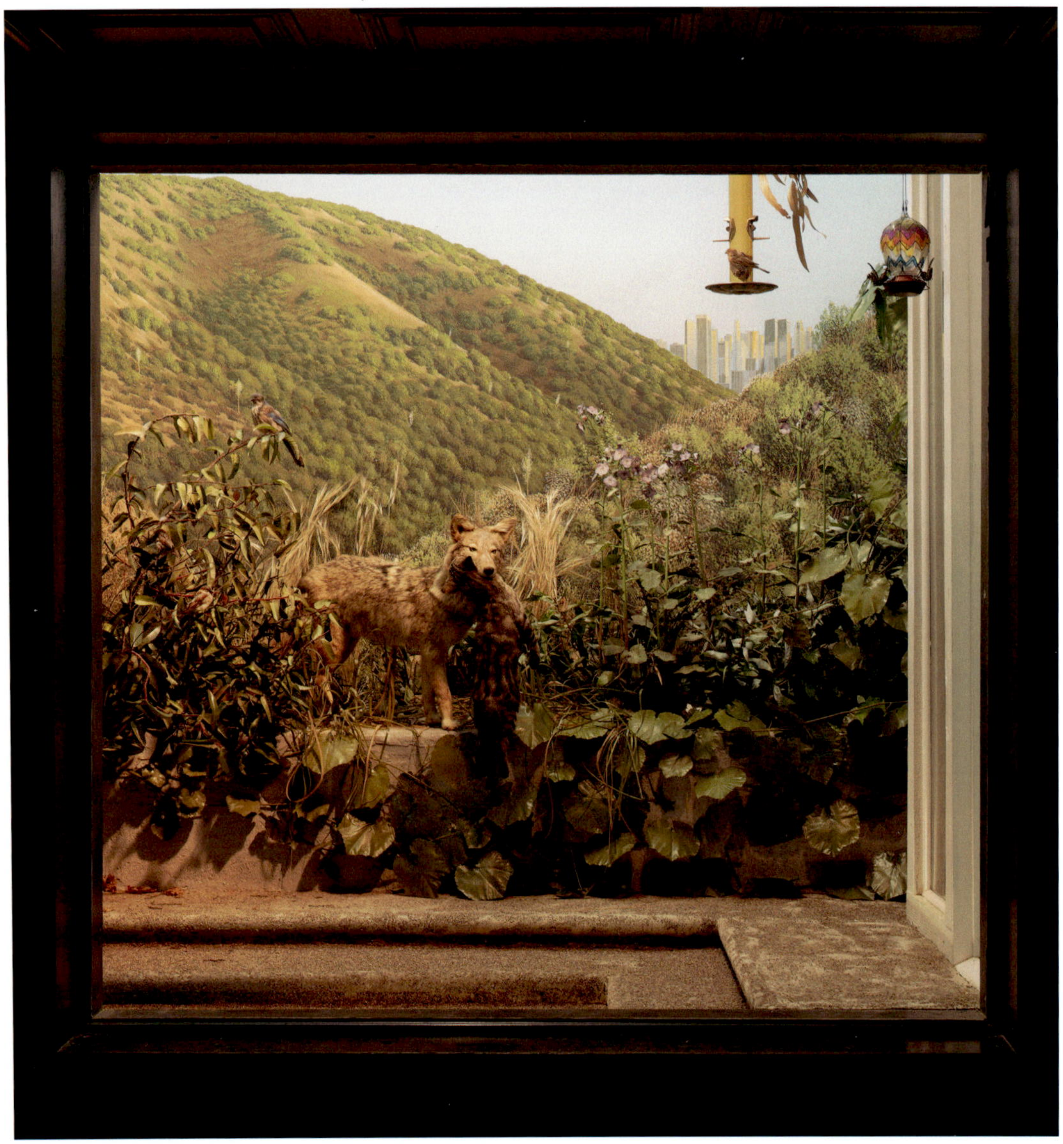

← NHM broke with tradition when it created this diorama of *L.A.'s Backyard*, which features introduced species thriving around a backyard pool with downtown Los Angeles in the distance (2021).

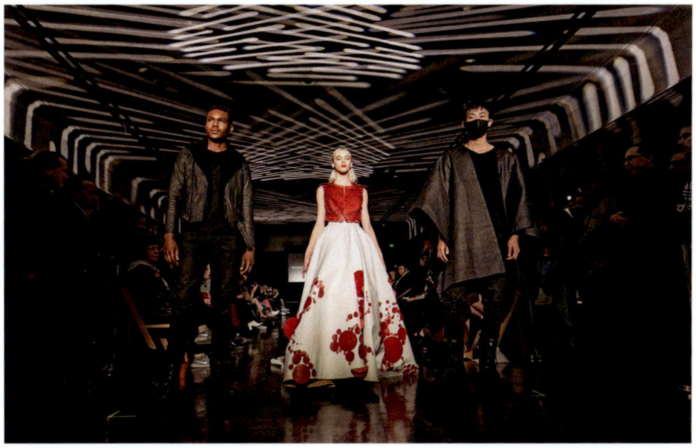

↑ A vegan fashion show took over the Hall of North American Mammals on Level 1, creating a curious contrast with the surrounding dioramas during Night of Ideas (2019).

↓ A sable antelope stands perfectly still for a visitor (2017).

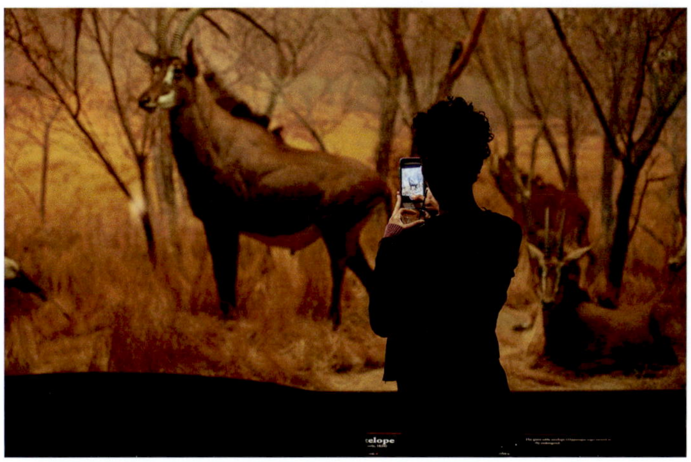

The diorama halls at NHM have held concerts, lively panel discussions about science and culture, and even a vegan fashion show. One may very well wonder what the taxidermied animals looking on might have thought, but it is not clear that these interventions change the way that people think about the animals. In fact, there is little research to show that dioramas actually alter visitors' attitudes about conservation. Often, they are just part of the scenery, a background for selfies.

So creating new meaning may require new experimental and even disruptive forms of dioramas. That is why NHM has invited outside artists to experiment with new forms in the halls. Some museums have decided to do away with dioramas because their meanings can be so refractory, so difficult to change, and sometimes problematic, as we've seen. But dioramas continue to attract large numbers of visitors, and their frequent appearances across various forms of popular culture attest to their continuing ability to entertain and communicate in surprising ways.

Perhaps we can learn from some of the ways that dioramas are used outside of the museum to inspire innovation inside it. By immersing us in a moment in the life of a nonhuman species—sometimes one that may soon disappear in the wild—these scenes enable us to travel to other spaces, times, bodies, and modes of life, even as they unsettle us with their uncanny combinations of reality, cultural perception, and aesthetic practice. For starters, diorama designers of the future could experiment with more human-altered landscapes, so-called novel ecosystems. They might also try out surprising spatial arrangements, virtual backgrounds, sounds, smells, animation, or holographs of animals along with real ones. The point, however, ought not to be putting a high-tech gloss on a traditional form but instead creating new forms. Some of these techniques could be used to create a narrative beyond a single scene. A series could build a narrative through a sequence in a hall, like a graphic novel.

In any case, diorama designers of the future will want to grapple seriously with a key insight that emerges from our contemporary encounter with dioramas. Nature, even and especially at its most authentic, never comes to us as unalloyed reality. It is always filtered through layers of cultural worldviews, social practices, historical memories, and anticipations of our future relationships with nonhuman species. What Lévi-Strauss concluded about totems can be applied to dioramas, too:

> Totem animals cease to be solely or principally feared, admired, or desired creatures: their perceptible reality makes apparent ideas and relations conceived by speculative thought on the basis of observed facts.[11]

Lévi-Strauss put thinking about nature through totems—that is, human beings comparing themselves with nonhuman beings, creating metaphors with animals and plants, speculating about differences and similarities, categorizing individuals and species—at the very origins of human intellect, language, and culture. "There is nothing archaic or remote about it," Lévi-Strauss concluded about totemism. And we would say the same about dioramas.

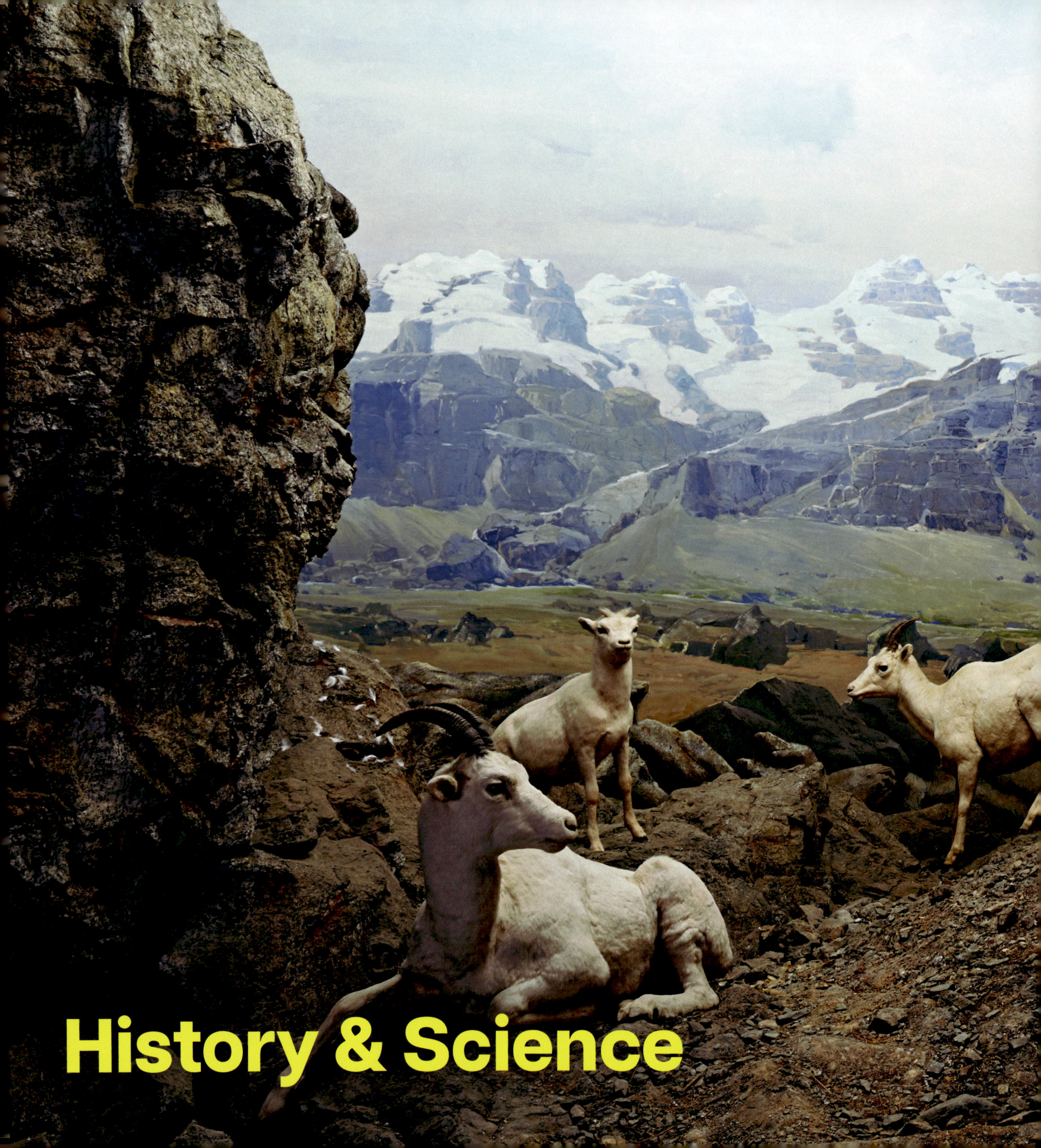

History & Science

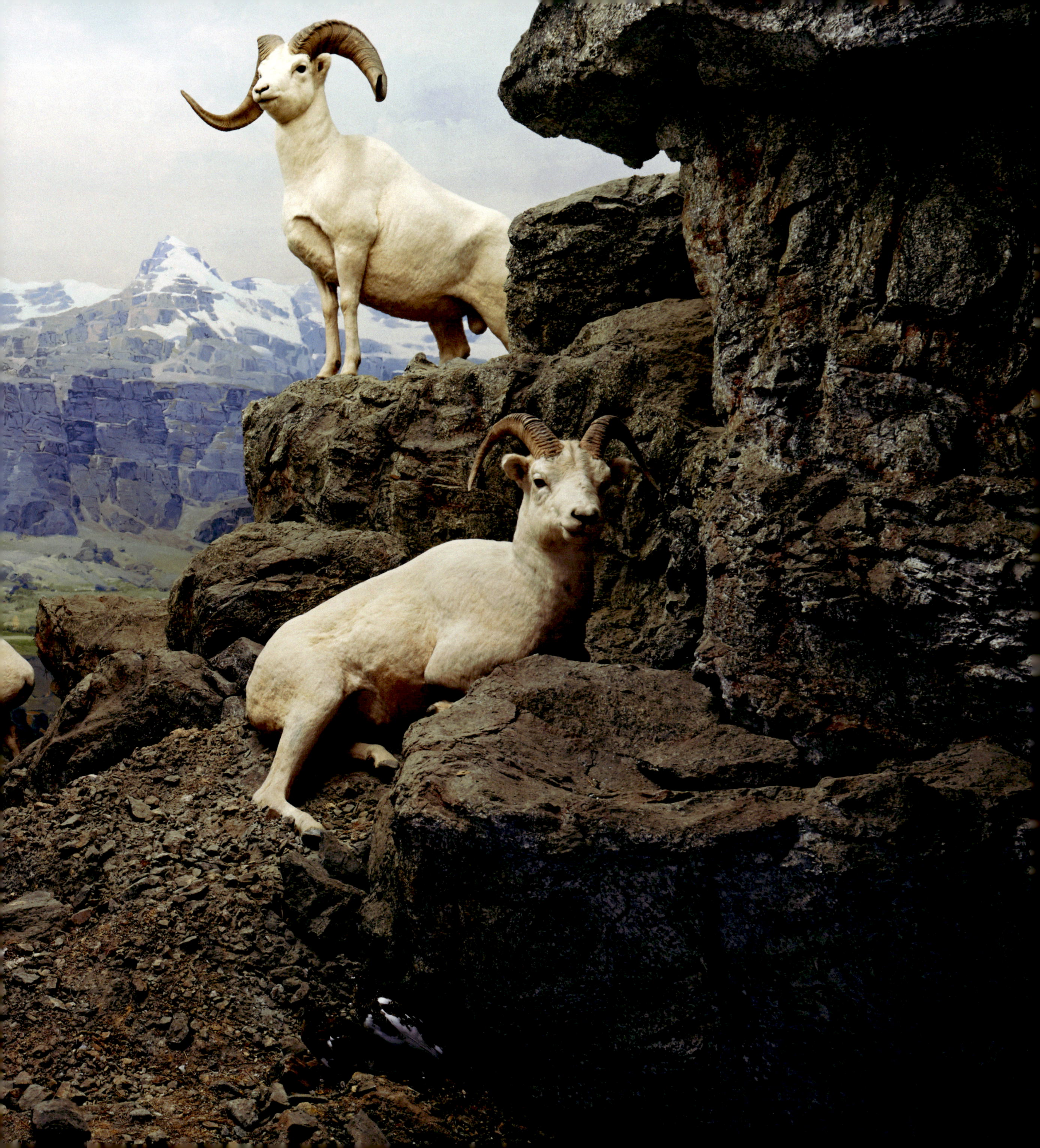

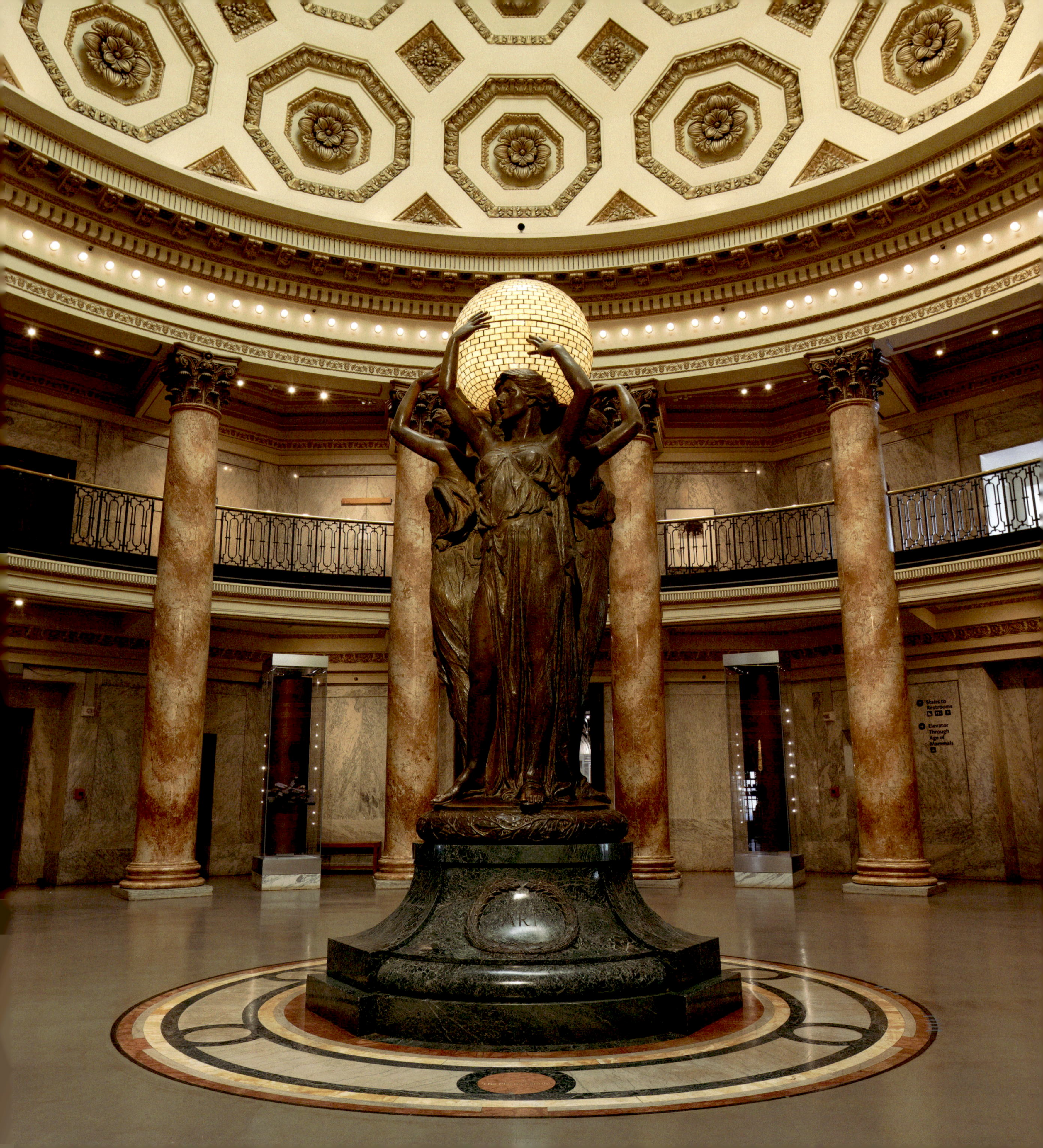
ART
Stairs to Restrooms
Elevator Through Age of Mammals

The First 100 Years

Matt Davis and Araceli Ramos

What we now know as the Natural History Museum first opened in 1913 as the Los Angeles County Museum of History, Science and Art. Befitting its original name, the museum had only three galleries: history, science, and art. There were no dioramas. Yet within only a few years, the first director of the museum, Frank Daggett, recognized the urgent need for educational diorama halls: "Owing to the fact that the big game animals of the territory are rapidly disappearing, it seems most important that our next move be made in the direction of securing groups for permanent exhibition in the future."[12] A major expansion was planned in phases to reorient the fledgling museum around four new diorama halls.

The creation of the new diorama halls was driven largely by three people; the first and most important was pioneering Taxidermist John Rowley. Rowley's key innovation was designing an entire hall specifically to display habitat groups. Still a relatively recent addition to exhibitions, dioramas at the time were often crammed into existing nooks in museums where bright windows caused distracting reflections. In 1916, Rowley opened the world's first mammal diorama hall at the California Academy of Sciences in San Francisco, to ecstatic reviews. "It is not venturesome to prophesy that the museum of the future will be planned almost exclusively

Previous spread
With a background painted in 1926, the *Dall's Thinhorn Sheep* group is one of the oldest dioramas at the museum. While the glaciers in the background appear frozen in time, the actual habitat recorded by the painting is rapidly warming. Dioramas, and the scientific specimens inside them, are valuable resources for understanding the changing world we live in (2014).

← Standing at the center of the Haaga Family Rotunda, Julia Bracken Wendt's 1914 sculpture *The Three Graces* pointed visitors to the original wings for history, science, and art. Only a few years after it opened, major expansions would reorient the museum away from the rotunda and recenter it around four new diorama halls.

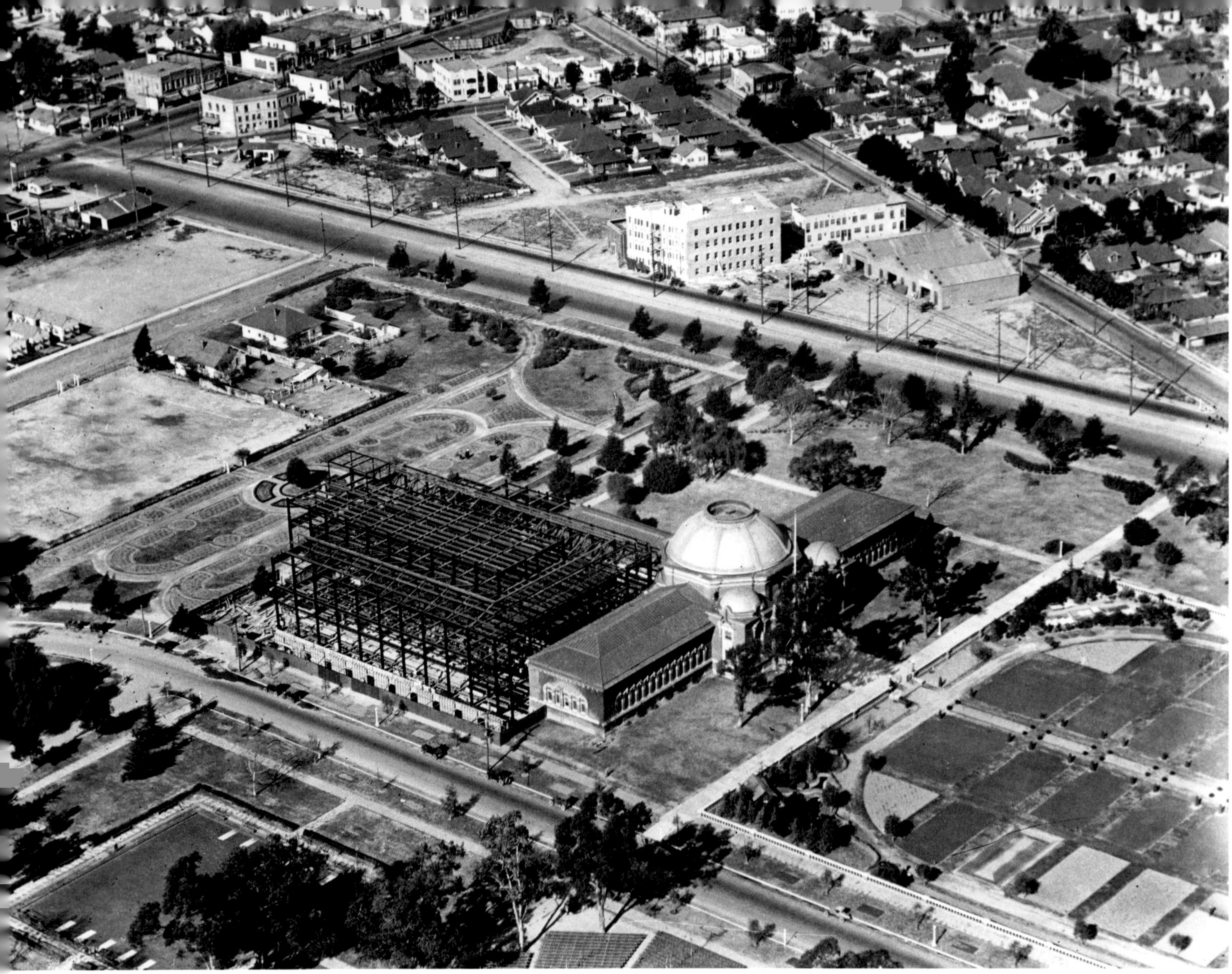

↑ The original building, opened in 1913, had only three galleries extending off a central rotunda: history, science, and art. The construction of the Unit I Addition, shown here in 1925, and later the Unit II Addition to the west (*left*) of that would increase the size of the museum tenfold and recenter it around four new diorama halls.

for habitat groups," exclaimed *Scientific American*.[13] The controlled perspectives in Rowley's purpose-built hall greatly increased the illusionistic realism of dioramas and amazed everyone, including museum leaders back in Los Angeles who were tired of boring cases stuffed with bones. "I can assure you that we do not desire to obtain any specimens for the purpose of exhibiting the same in the old glass case manner," wrote the acting director of the museum to the California Fish and Game Commission. "We are going to spend a great deal of money, and we feel that we can at least equal the San Francisco Academy's exhibit—and hope to improve upon it."[14]

The others most responsible for NHM's early expansion were the two men who would fund, fill, and lend their names to the museum's first diorama halls: the Honorable Reuben F. McClellan and Leslie Simson. They were

→ John Rowley, pioneering taxidermist and first leader of NHM's dioramas, sits (*second from left*) with his staff in 1922.

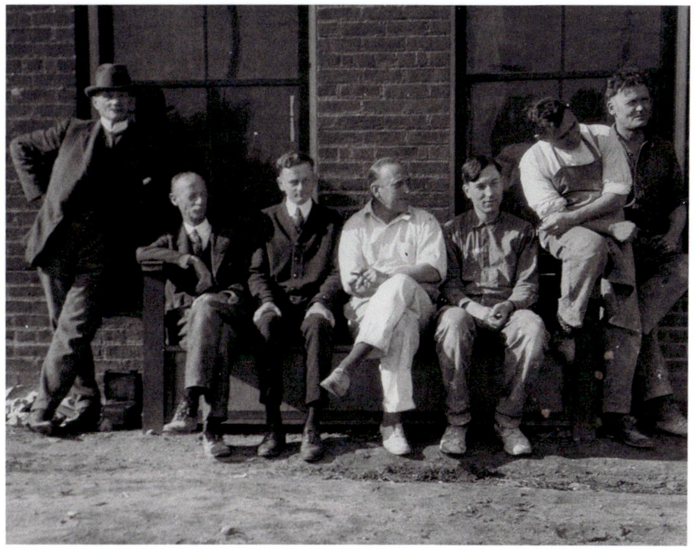

→→ Taxidermist John Rowley's key innovation was designing an entire hall solely for habitat group dioramas where controlled perspectives and lighting made the scenes appear even more immersive and lifelike. His plan for the ideal hall, published in his 1925 book *Taxidermy and Museum Exhibition*, is almost identical to the halls he designed for NHM.

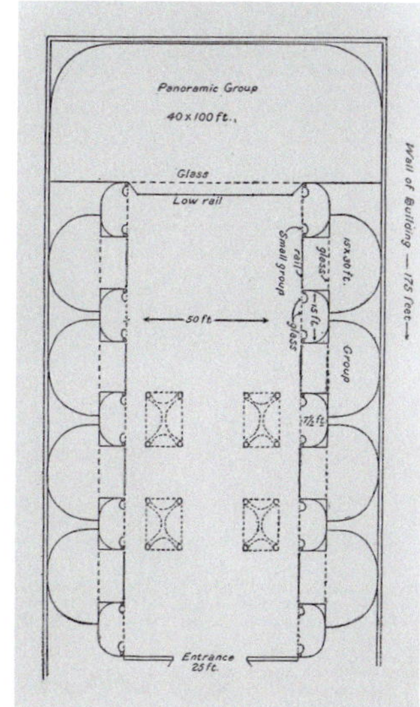

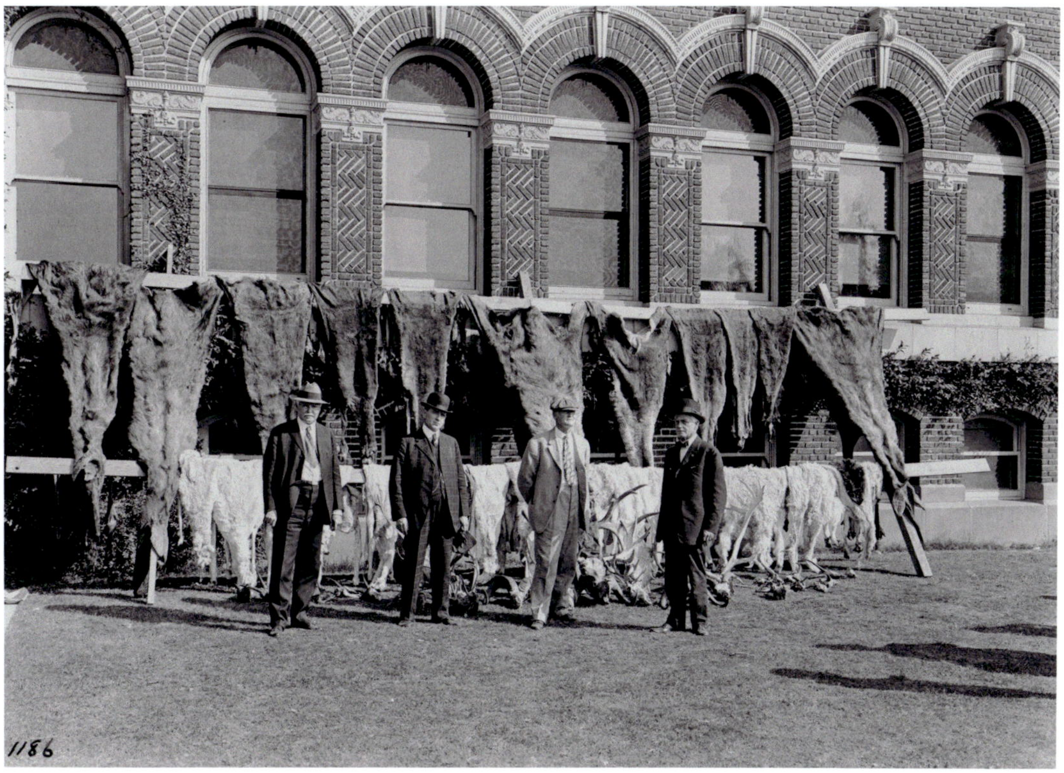

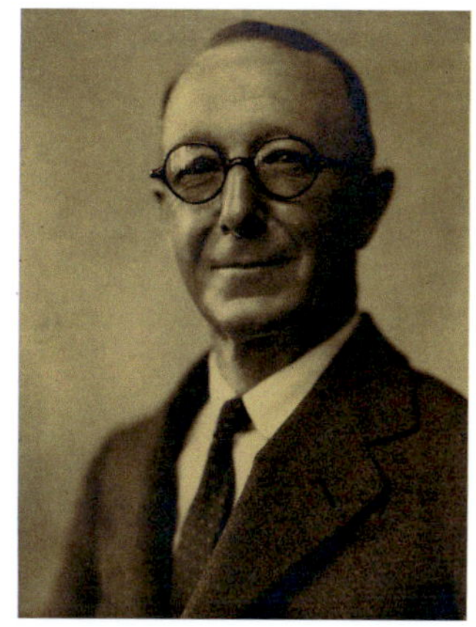

↑ Leslie Simson was a wealthy mining engineer from California who personally hunted all the animals for NHM's original Simson Hall of African Mammals, now the Mr. and Mrs. Samuel Oschin and Family Hall of African Mammals (c. 1930).

↑ County supervisor and expedition leader Reuben F. McClellan (*right*) and Museum Director Frank Daggett (*left*) pose with two others in front of the original museum building with pelts of caribou, moose, mountain goats, and bighorn sheep collected for the dioramas on the 1919 Alaska Research Expedition.

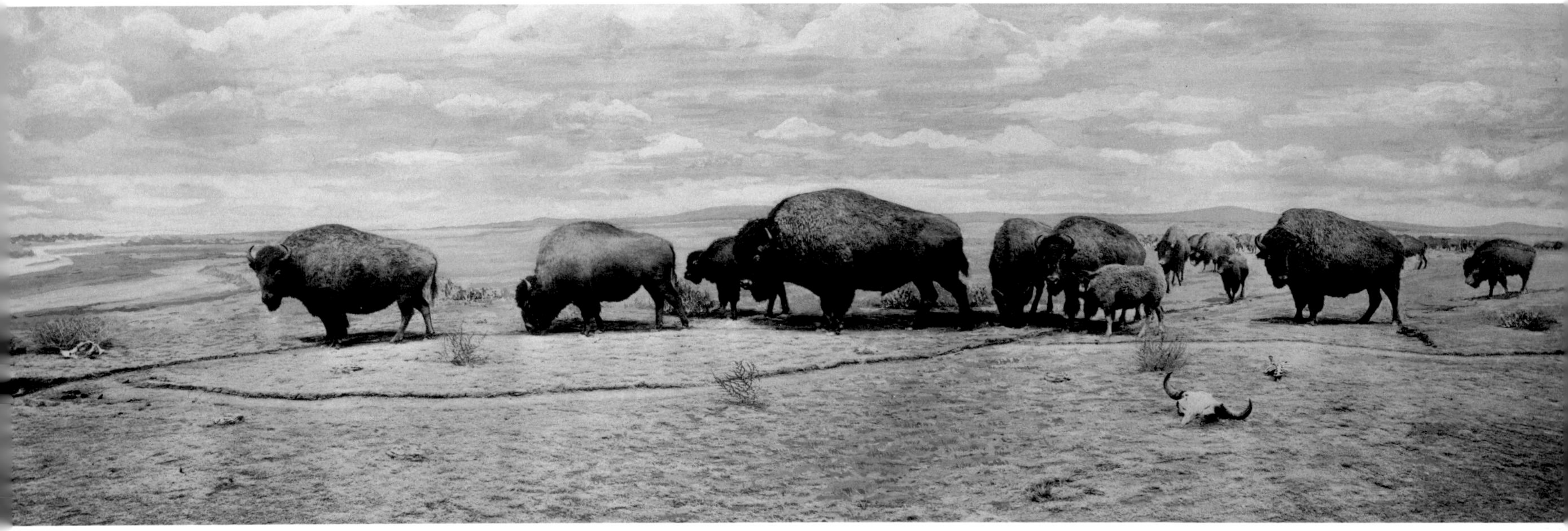

similar in many ways. Both were rugged outdoorsmen and experienced big game hunters who made fortunes in mining: McClellan in Alaska, Simson in South Africa. Sitting on the museum's Board of Governors and L.A. County's Board of Supervisors, McClellan championed the museum and the idea of creating inspiring diorama halls for the growing populace of Los Angeles. In 1919 he personally led the museum's first major expedition—a trip to Alaska to hunt big game for the North American Hall.

Unlike McClellan, Simson was a reserved lifelong bachelor who preferred the plains of Africa to bustling cities like Los Angeles. Despite his deserved reputation for hunting prowess, Simson was incredibly modest and private. He rarely promoted his adventures, and he refused to let the museum laud his world record number of lion kills in press releases. Friends with Rowley, Simson offered to collect an entire hall's worth of African animals at his own expense as long as Rowley could get it built. Unable to secure funding in the Bay Area for a new hall, Rowley took Simson's offer to Los Angeles, where it was eagerly accepted. At NHM, Rowley perfected his vision for the ideal diorama hall. Unlike San Francisco's sunlit hall, each NHM diorama would be lit by separate electric lights, giving him complete freedom to set scenes anytime from dawn till dusk.

The Unit I Addition opened in November 1925, more than quintupling the exhibition space of the original

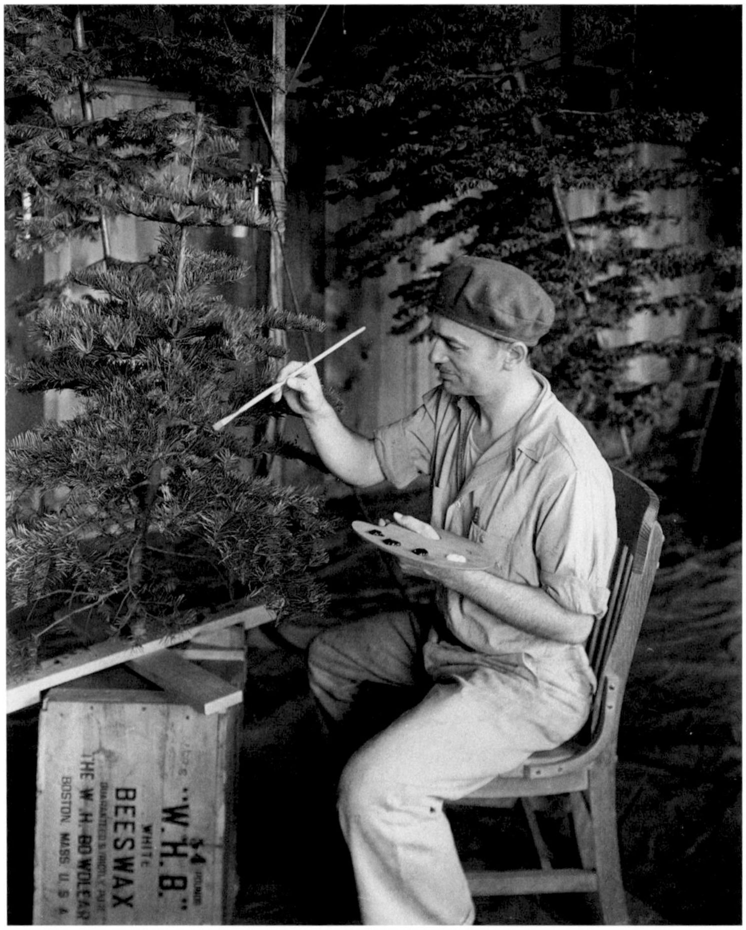

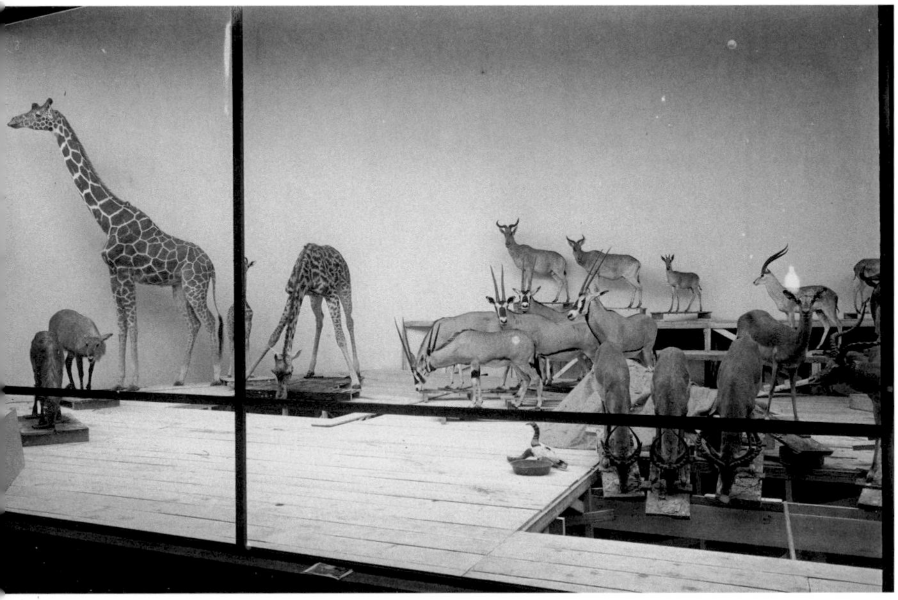
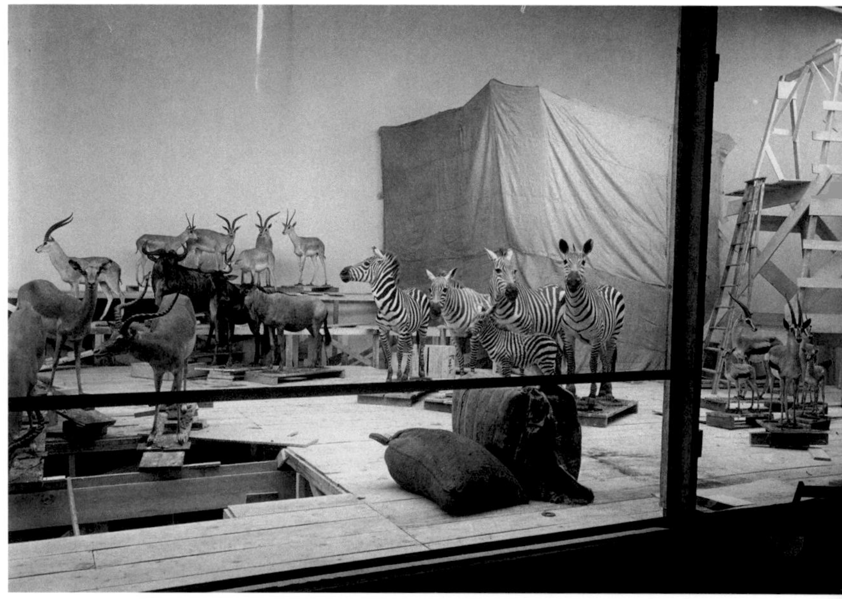

↖ The McClellan Hall of North American Mammals opened in 1925 with the *Bison* diorama. The hall was named after L.A. County supervisor Reuben F. McClellan, who collected many of the animals on display and championed the museum.

↑ Since Rowley refused to hand over the plans for the diorama halls upon being fired, Lincoln had to figure out how to arrange the hundreds of animals Rowley had already mounted into realistic scenes. These photos show an early layout test for the first *Watering Hole* diorama (1930).

← In 1930, young Foreground Artist Melville Lincoln was given the task of finishing the Simson Hall of African Mammals in record time (now the Mr. and Mrs. Samuel Oschin and Family Hall of African Mammals). He would go on to lead NHM's diorama program for the next three decades (1931).

museum. Visitors poured into the McClellan Hall of North American Mammals to see the first habitat diorama in Los Angeles, a group of bison shot by McClellan himself. Museums as far away as New York City wrote to obtain pictures of the group that had "occasioned so much favorable comment," intending to inspire diorama halls of their own.[15]

But trouble was already brewing. As Rowley worked to fill the remaining diorama bays, he got into a feud with his assistant, Taxidermist Joseph P. Herring, over methods and artistic direction for the halls. To keep the peace, Director William Bryan divided the new halls between them. Rowley worked only on Africa, while Herring got North America. But this truce was not enough, and a grand jury was convened to investigate the museum's diorama program. Its report, which made front-page news, lambasted Bryan for tolerating the feud and allowing the dioramas to run so far behind schedule and over budget. The fallout was swift; Bryan fled to Europe to wait out the public relations disaster; Rowley and Herring were fired.

Still, civic leaders expected the Unit II Addition and its new African diorama hall to open in fall 1930. Once back from Europe, Bryan hired Background Artist Frank MacKenzie and promoted Foreground Artist Melville Lincoln to oversee the dioramas. MacKenzie arrived at the museum ready to paint but was flabbergasted that nothing had been prepared or planned yet. In Rowley's final act of defiance, he had refused to hand over all his notes and plans for the dioramas. Lincoln and MacKenzie were left to puzzle out how to arrange the 214 animals Rowley had already mounted into accurate habitat groups. Lincoln pored over botany books to find plants in the local hills that looked African enough to use for the diorama foregrounds; he didn't get everything right. The baby giraffe in the *Watering Hole* group was a different subspecies than its "father" standing nearby. The kudu were put in a habitat more suitable for wildebeests and vice versa. MacKenzie had served in the military in colonial Africa but wasn't an animal expert, and Lincoln had never even seen the continent. Though their scientific accuracy can be questioned, their speed was incredible. For comparison, it took over 30 years to complete the Akeley Hall of African Mammals at the American Museum of Natural History in New York City. Lincoln and MacKenzie had finished all 17 dioramas in the Simson Hall of African Mammals in only seven months, just in time for the opening of the Unit II Addition. Almost 50,000 people came to see the first African diorama hall in the country.

Although the two new museum units were now open, not all the diorama bays were filled. The North American Hall still had many empty habitats, and the halls on Level 2 were completely empty. Lincoln and his team worked steadily to fill in the dioramas, but museum leadership became increasingly frustrated with Lincoln's disdain for oversight and the quality of his work. Shouldn't the scientific curators of the museum be in charge of the habitats, not an artist? In 1949, they asked Ralph Morrill of the Peabody Museum of Natural History at Yale University to give an independent review of NHM's diorama halls. Morrill carefully examined each diorama before delivering his withering report. "It is my opinion that the habitat groups recently completed are technically not of sufficient quality to justify continuation of the present program," Morrill wrote. "In most cases, perspective is faulty and the lighting is primitive and inadequate. The backgrounds are on the whole, very badly executed. The accessories [foreground] work is old and dusty."[16] Lincoln was devastated by the review and urged to retire. In his defense, it was true that the museum dedicated less money and fewer staff to its dioramas than many peer institutions, but that was about to change.

Spurned by Morrill's scathing review, the museum reinvigorated its diorama program by launching a series of ambitious expeditions in the 1950s and '60s led and funded by affluent Los Angeles families. The Logans brought back muskoxen, caribou, and walruses from the frozen North; the Garlands and Cheneys hunted through East Africa; and the Machrises went everywhere in between to gather specimens for the museum's collections and dioramas. These midcentury expeditions were a stark departure from the early patronage of lone sportsmen like Leslie Simson, who would ship back whichever large animals he happened to be hunting on safari in Africa. These trips were extensively planned and included a contingent of scientists, taxidermists, and artists to collect everything that would go into an animal's habitat. Hunting rhinos and hippos was a way to get patrons and the public excited about expeditions to collect small birds, fish, and insects for biodiversity research. This period reached its peak with the opening of the new *Watering Hole* group in 1966. It was the largest diorama in the world and the only one to feature a whole family of African elephants.

Despite this high point, dioramas were already starting to fall out of favor. They were still popular with visitors, and several museums continued to install them, but few imagined "the museum of the future" being built completely around habitat groups anymore. The increasing

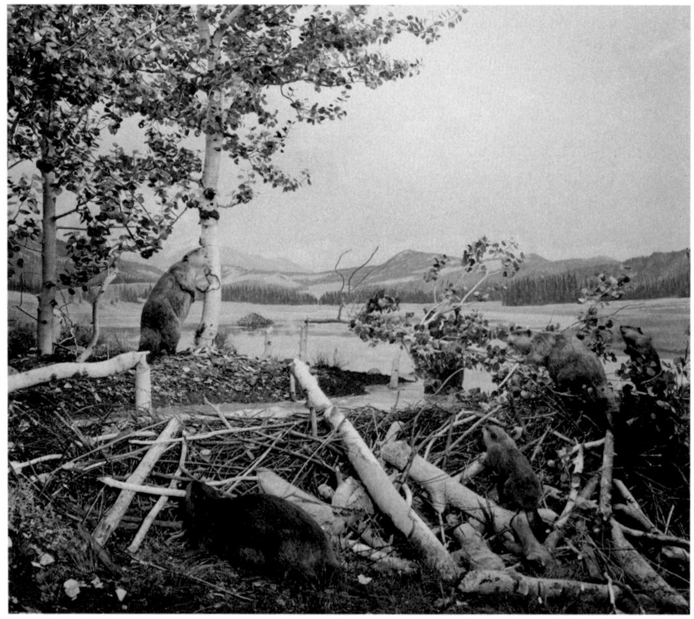

↑ Yale's Ralph Morrill singled out the *North American Beaver* group in his devastating review of NHM's dioramas, writing that the water looks like "sour milk" and that the "dam is terrible and would make any self-respecting beaver hang his head in shame."[17]

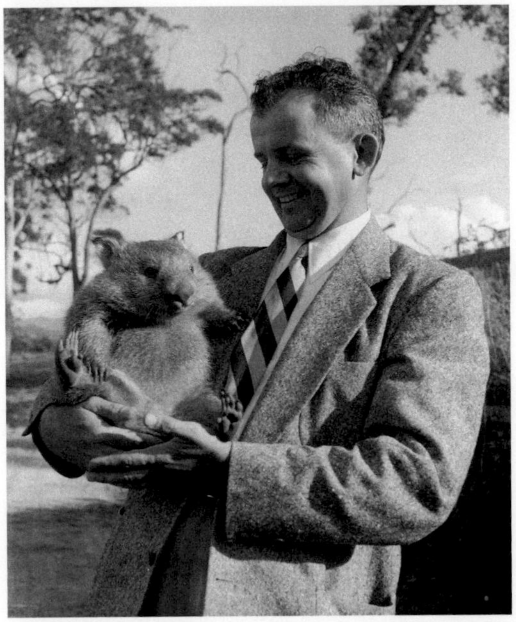

← Curator of Birds and Mammals Kenneth E. Stager holds a new wombat friend on the 1954 Davidson Australian Expedition.

↓ Members of the 1956 Garland East African Expedition pose triumphantly with one of the black rhinos they would bring back to the museum. *Left to right:* Senior Curator of Habitats Melville Lincoln, Good Samaritan Hospital physician Dr. Henry Morgan, Taxidermist George Adams, and expedition sponsor John J. Garland. Today, black rhinos are classified as critically endangered and most would find killing them repulsive, but 70 years ago, when Dr. Morgan shot this male for the museum, the species was over 30 times more common. Newspapers praised the hunters for filling in a diorama "long-wanted by the county."[18]

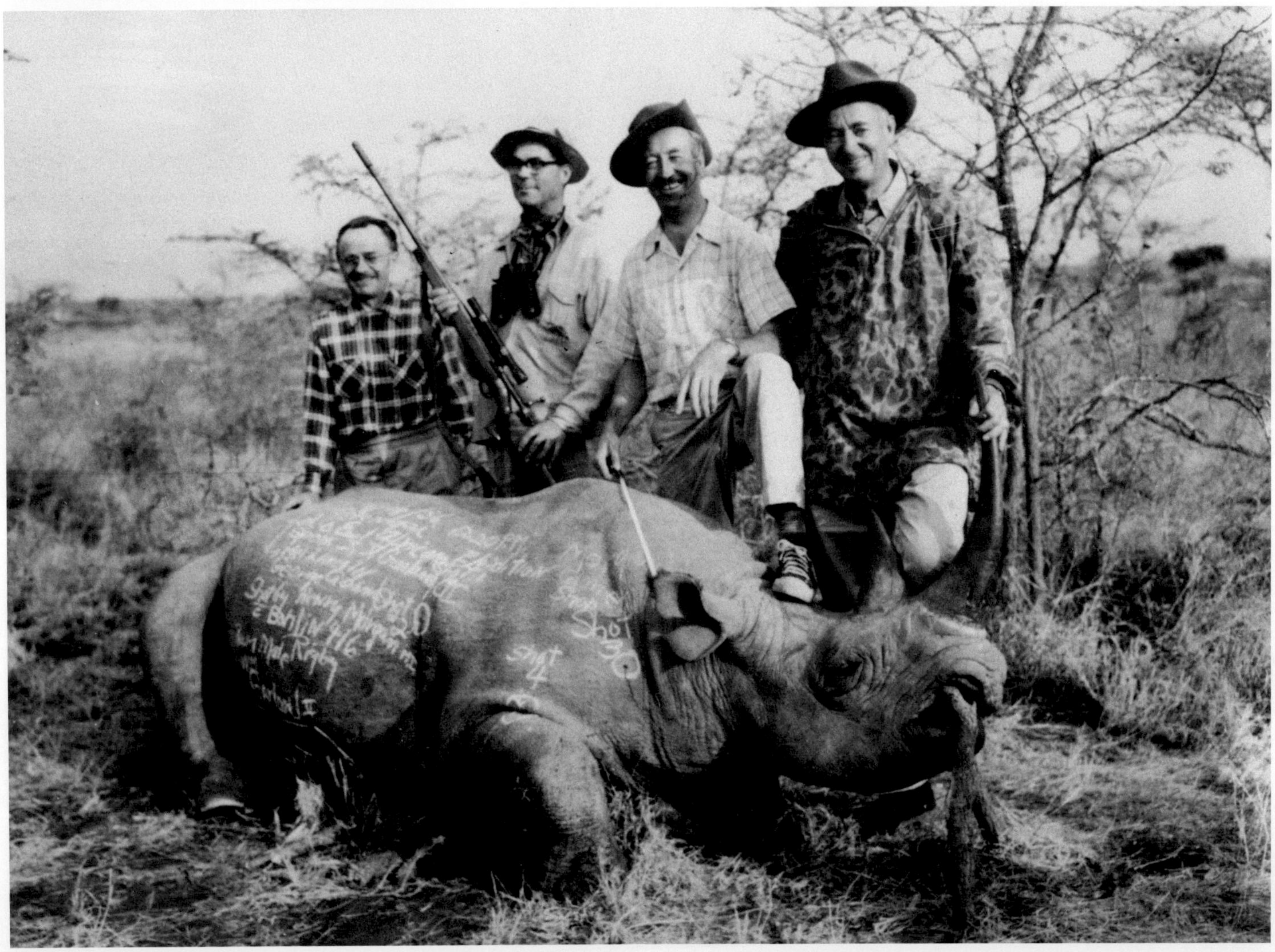

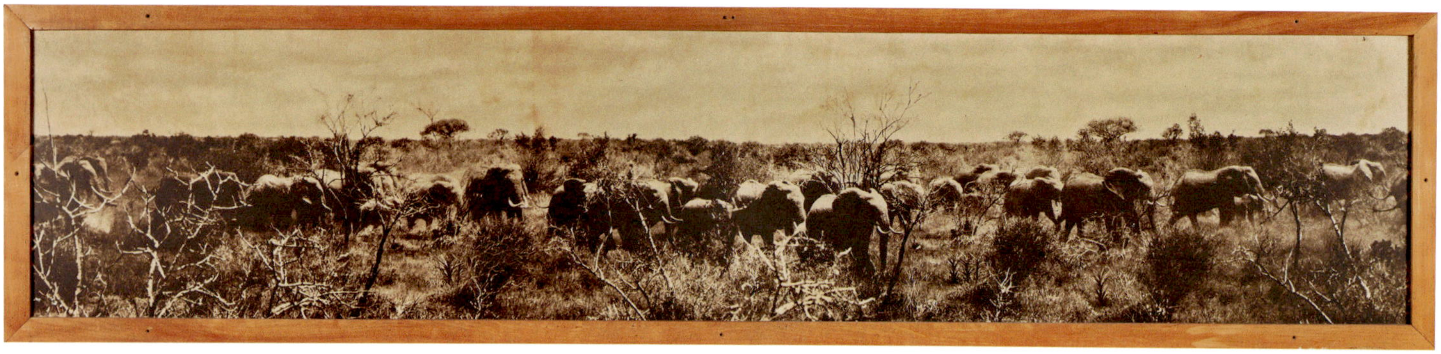

↑ This large panorama of elephants stampeding through the Kenyan bush on the 1963 Knudsen-Machris Expedition has hung in the Taxidermy Studio for decades.

→ On the 1969 Logan Expedition, a hunter scans the mountains of British Columbia, Canada, for the elusive Stone's thinhorn sheep.

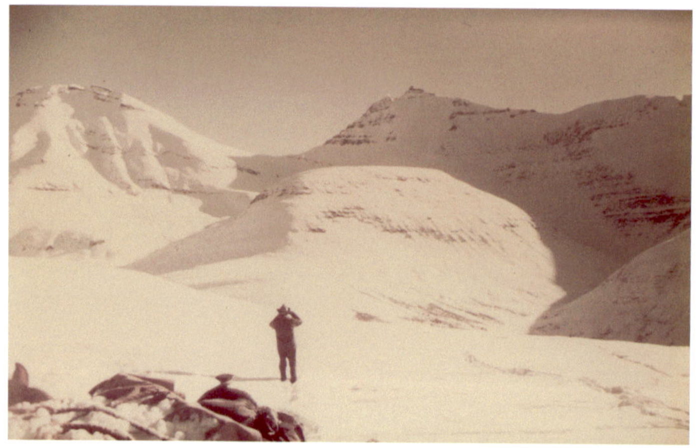

↓ The museum's second *Eland* group was collected on the 1956 Garland East African Expedition (2021).

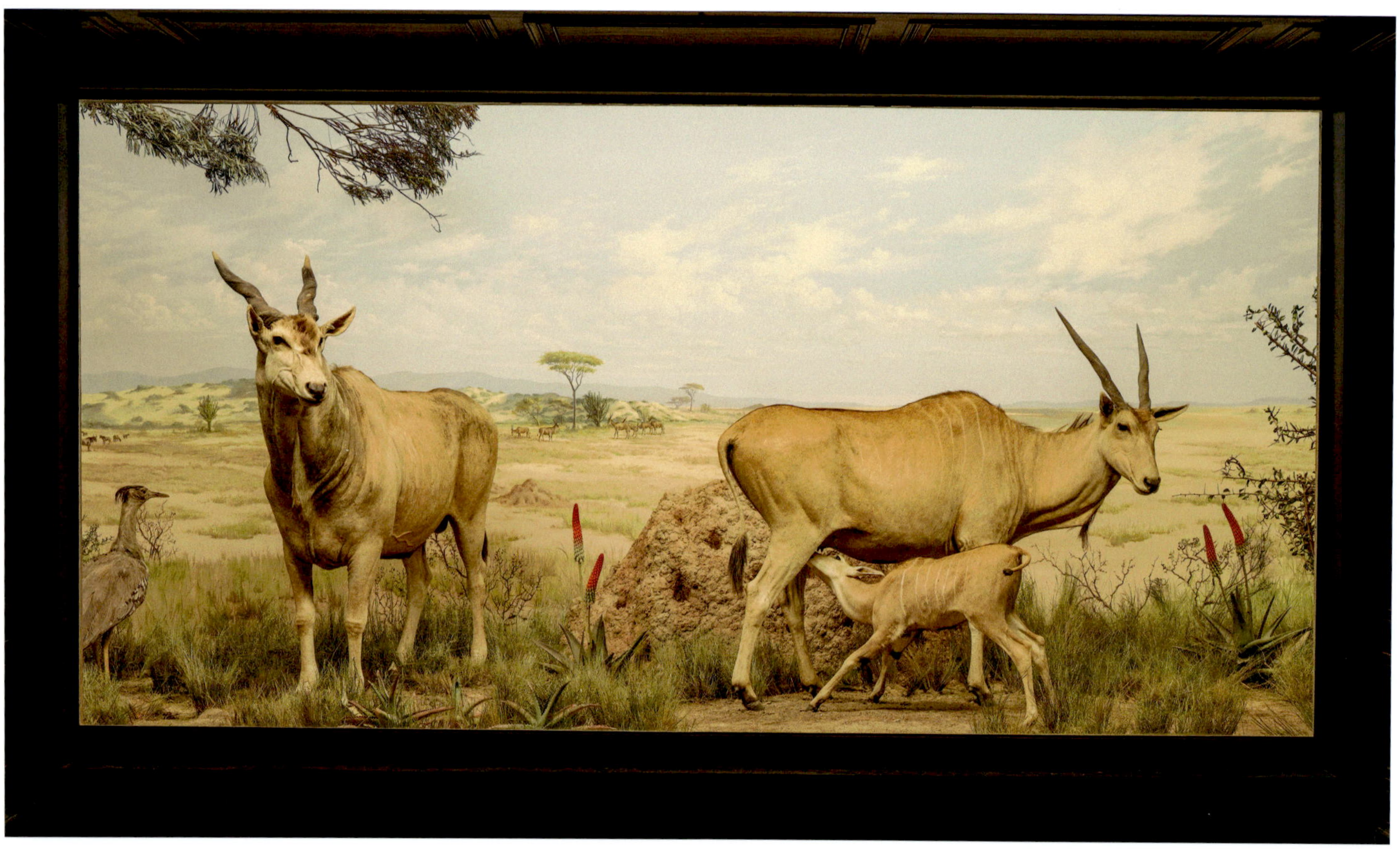

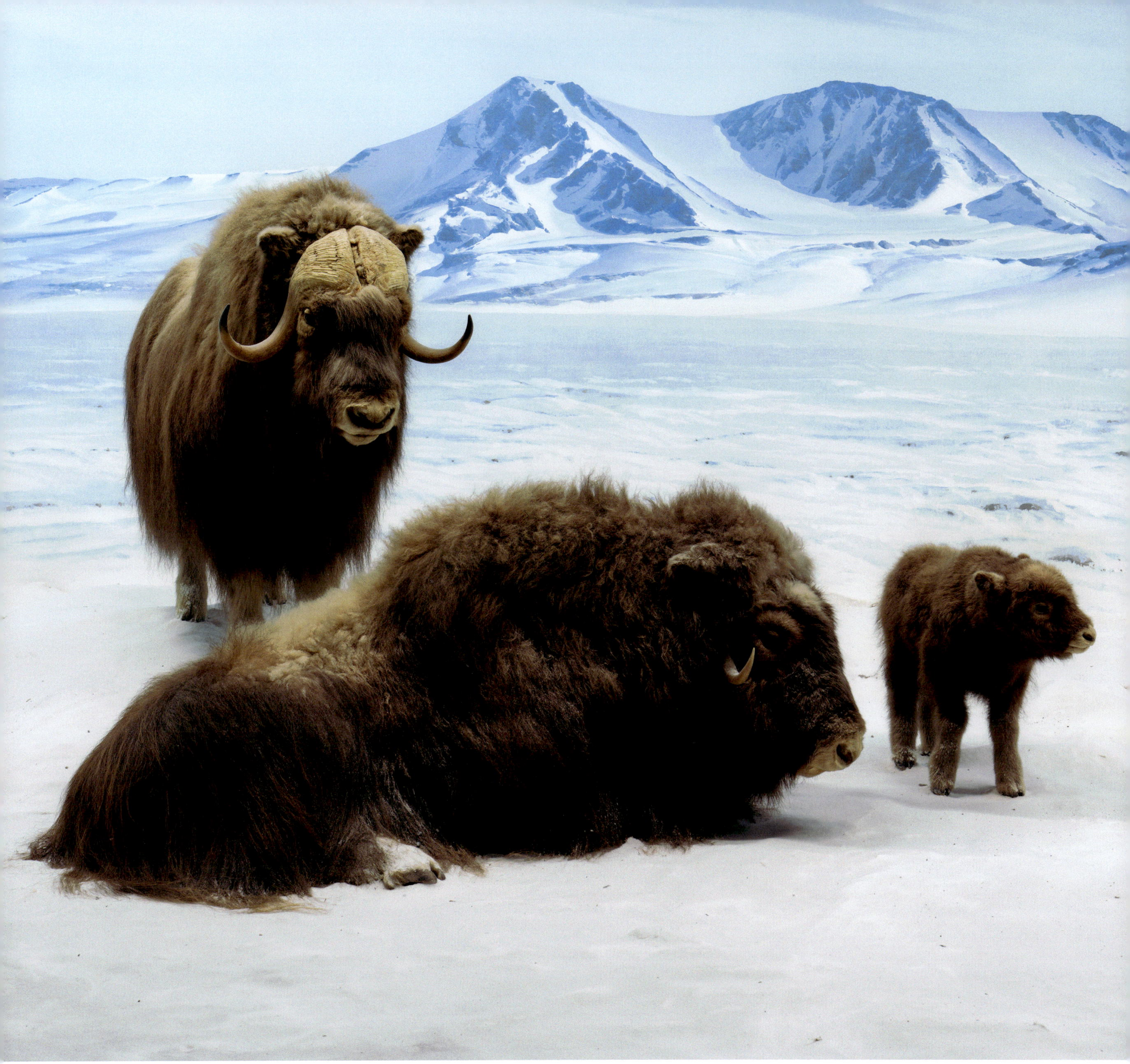

↑ The *Muskox* group was collected on the 1960 Logan Arctic Expedition (2021).

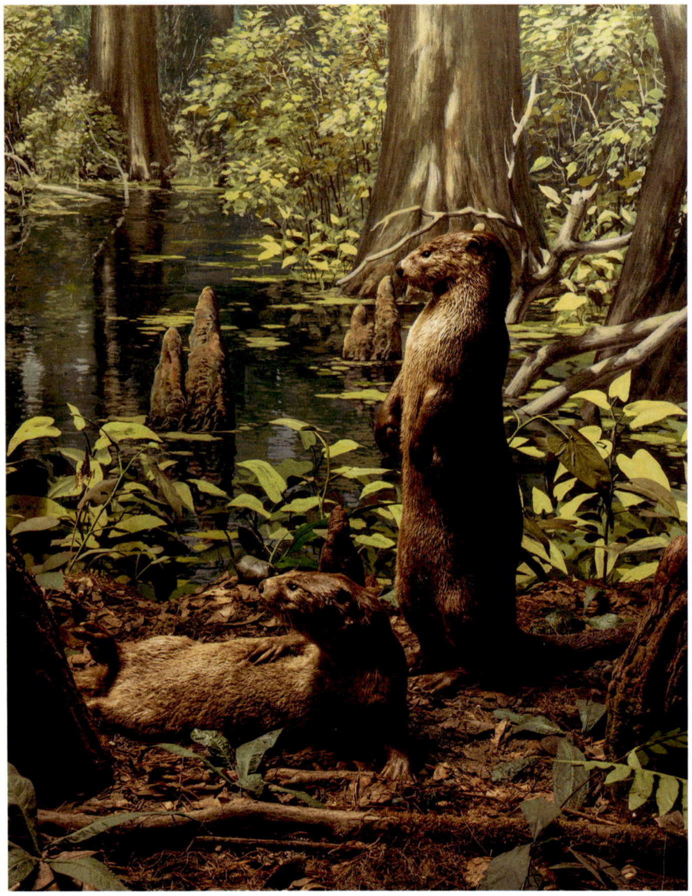

↑ The *North American River Otter* diorama, opened in 1974, reflected a general trend featuring smaller mammals and lush foregrounds (2021).

availability of nature documentaries in color and a growing distaste for hunting made dioramas seem like an expensive and cruelly outdated way to educate the public. Adventurous patrons exploring the wilds began to sound more like rich colonialists who used the museum's auspices to make hunting rare game a tax write-off. Director Herbert Friedmann posed the dilemma in 1964: "How then are we to justify habitat groups? What do they convey to the visitor that cannot be communicated as readily or as directly by other types of thematic or topical exhibits?"[19]

One answer was the irreplaceable sense of wonder and discovery dioramas gave visitors, as if they were actually out exploring the wild. Other exhibits were exactingly prescriptive: "Look here," "Pull this," "Know that." In a world of rigid schedules and vanishing outdoor free time, dioramas were anomalous, open-ended invitations to exploration: "Look where you will, discover what you can, tell the story that you wish." Before 2024 the diorama labels didn't even list all the species that could be found in each group; every animal hidden behind a foreground leaf or rock was a surprise. The defenders of dioramas in the museum considered this sense of discovery reason enough to continue filling in the unfinished halls through the 1970s and '80s. Yet the habitats during this period were noticeably different, both in their general appearance and in their higher quality.

NHM began the 1970s with perhaps the greatest concentration of artistic talent in the dioramas' history: experienced Hollywood artist Duncan Spencer painted beautiful, hyperrealistic backgrounds in record time, while George Adams and Paul Hill continued sculpting incredibly lifelike taxidermy. New Chief of Habitats James D. Olson completed the team by bringing in skilled model makers who created leafy, lushly detailed foregrounds unimaginable even a decade before. The animals themselves were also different, often smaller and more common, like otters or opossums. By 1980, Olson was publicly announcing, "The days of grandiose museum field expeditions to exotic places, expressly for hunting purposes, have long vanished."[20] The larger mammals installed in the halls during this time, like the maned wolf, snow leopard, and okapi, were not hunted in the wild at all but instead were donated by zoos after they died.

Though dioramas are a meaningful way to use these "salvage" zoo specimens, the dichotomy between hunter and conservationist was always false. Wealthy patrons Maurice and Paquita Machris personally shot elephants, tigers, and chimpanzees for the museum. But the Machrises also used their fortune and connections

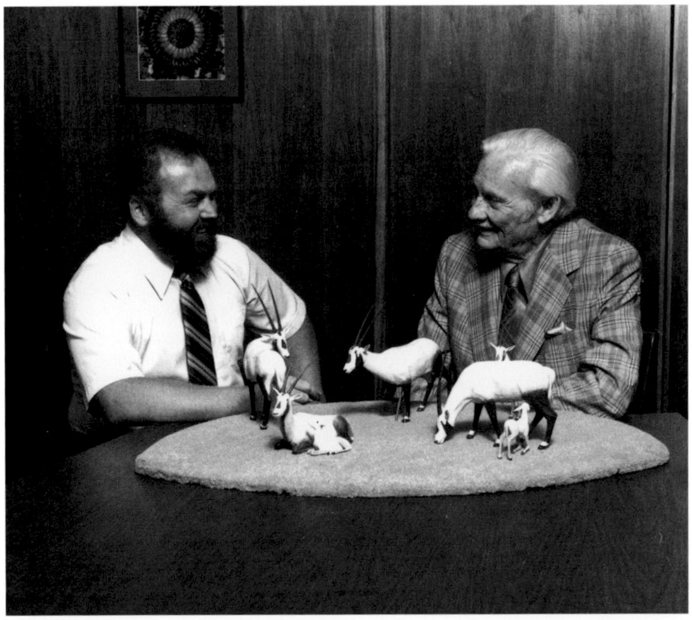

← Longtime museum donor Maurice Machris (*right*) discusses a scale model for the world's first Arabian oryx diorama with Chief of Habitats James D. Olson (*left*). The oryx were donated by zoos after they died rather than shot in the wild. Machris and his wife, Paquita, helped fund and organize a captive breeding program to save the endangered oryx from extinction (1980).

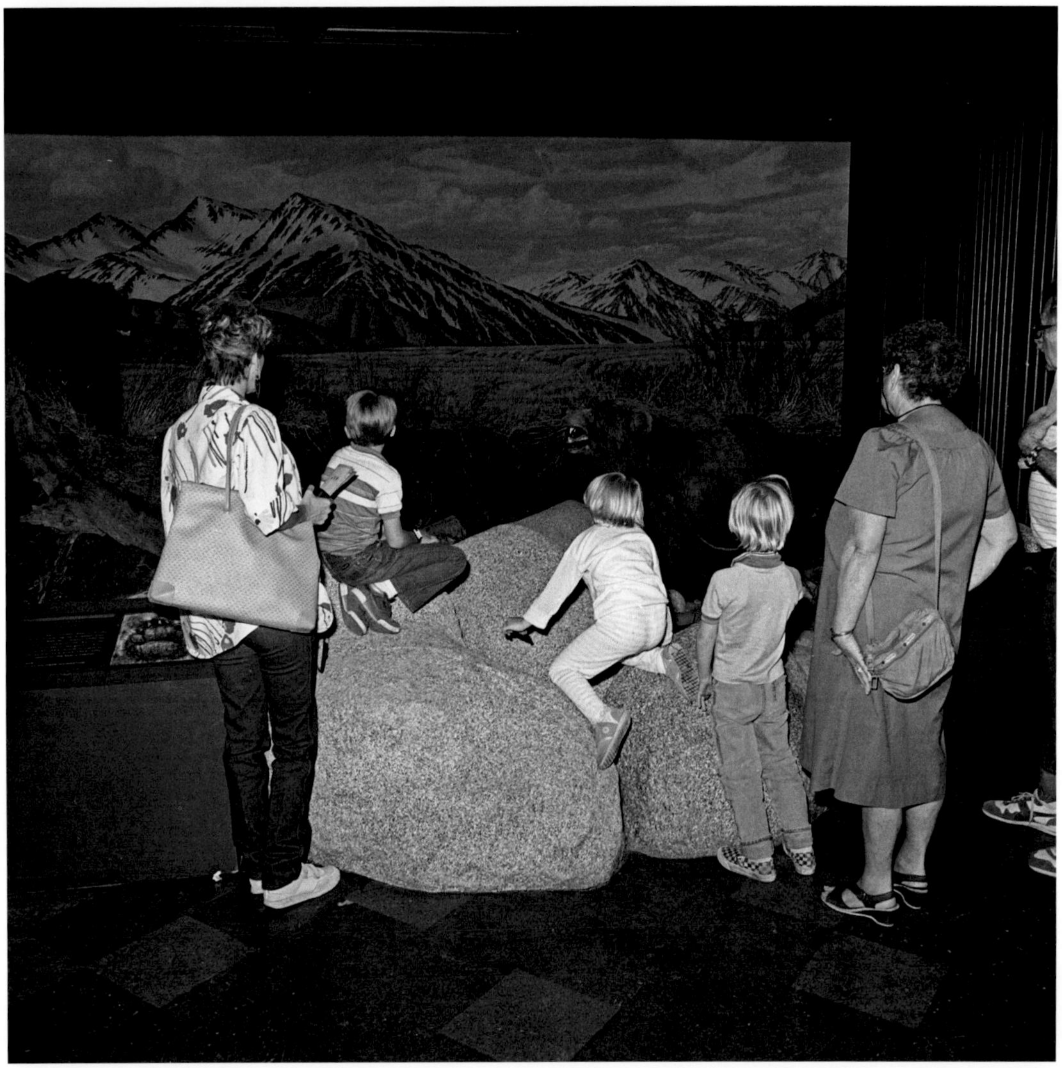

← NHM began experimenting with more immersive dioramas in 1986 by installing a new *Kodiak Bear* group with animatronics, a habitat that spilled out past the glass into the hall, and labels with touchable pelts and footprints (1986).

to start a captive breeding program for Arabian oryx that helped save this culturally important species from near certain extinction. The diorama the Machrises funded to celebrate the program's success, the first Arabian oryx group in the world, used only salvage specimens from zoos.

Dioramas changed in other ways too. Through the 1980s and '90s, the museum experimented with the rigid form of the habitat group. Habitats spilled through the fourth wall of the diorama glass, and real water gurgled past hippos, while animatronic elephant ears flapped back and forth. The innovations were meant to grab visitors' attention, but they proved a bit too inviting for patrons, who often clambered inside the dioramas and accidently trampled taxidermy mounts underfoot. The high-maintenance animatronics were eventually removed. During a major renovation in 2006, the groups were returned to their traditional positions behind glass. With fewer and fewer staff dedicated to dioramas, the taxidermist, assisted by a large team of volunteers, became responsible for the design, model making, and maintenance of the habitats. Since the renovations, the museum has installed several new dioramas, like the *Jaguar*, *Margay*, and *Island Fox*, but most recent work has focused on increasing biodiversity in existing dioramas. Scores of small birds, mammals, reptiles, and insects have been tucked into the foregrounds of habitats as if they were there all along.

As they turn 100, NHM's diorama halls are now some of the oldest in the country, yet their history is not over. In 2024 the museum reopened the former Exotic Mammal Hall to reveal dioramas that many visitors never knew existed; the hall was never finished and had remained shuttered for decades. To complete the hall, NHM invited outside artists to fill in several groups in nontraditional styles to provide different perspectives on what dioramas could be. How the halls evolve next has yet to be written, but if their century-long history is any indication, they will not remain still.

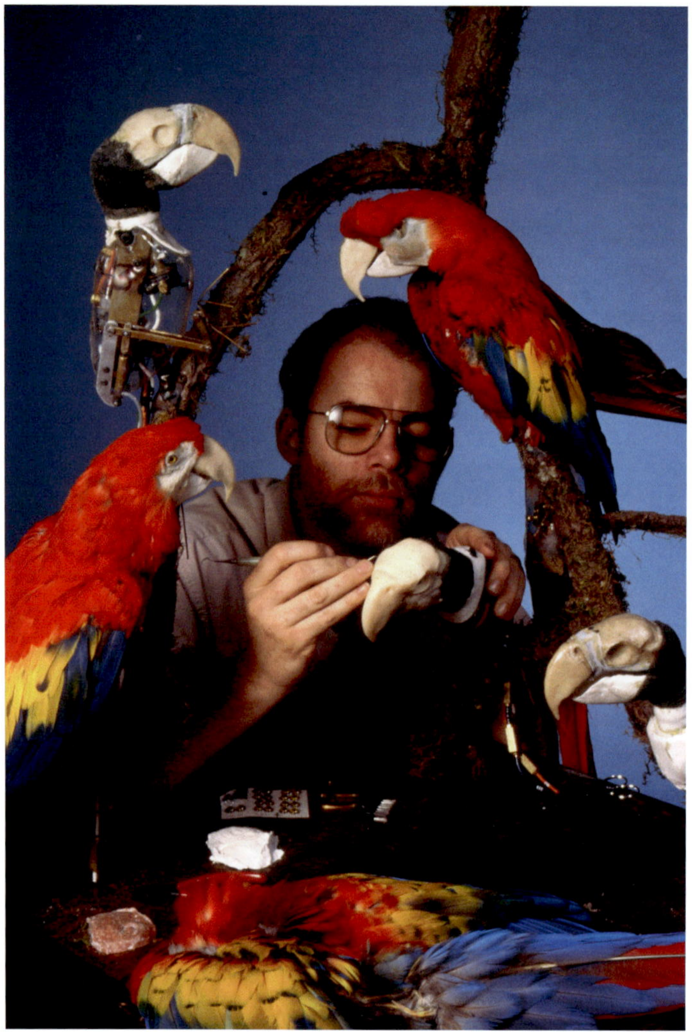

↑ During the early 1990s, NHM started installing more immersive and animated habitat groups throughout the diorama halls and the nearby Ralph W. Schreiber Hall of Birds, where these animatronic macaws being made by Taxidermist Tim Bovard would "live" (c. 1990).

→ After a 2006 renovation that removed the animatronics and returned the halls to a more traditional aesthetic, the *Watering Hole* group became the only remaining diorama at NHM that breaks the fourth wall by extending out past the threshold and into the hall (2021).

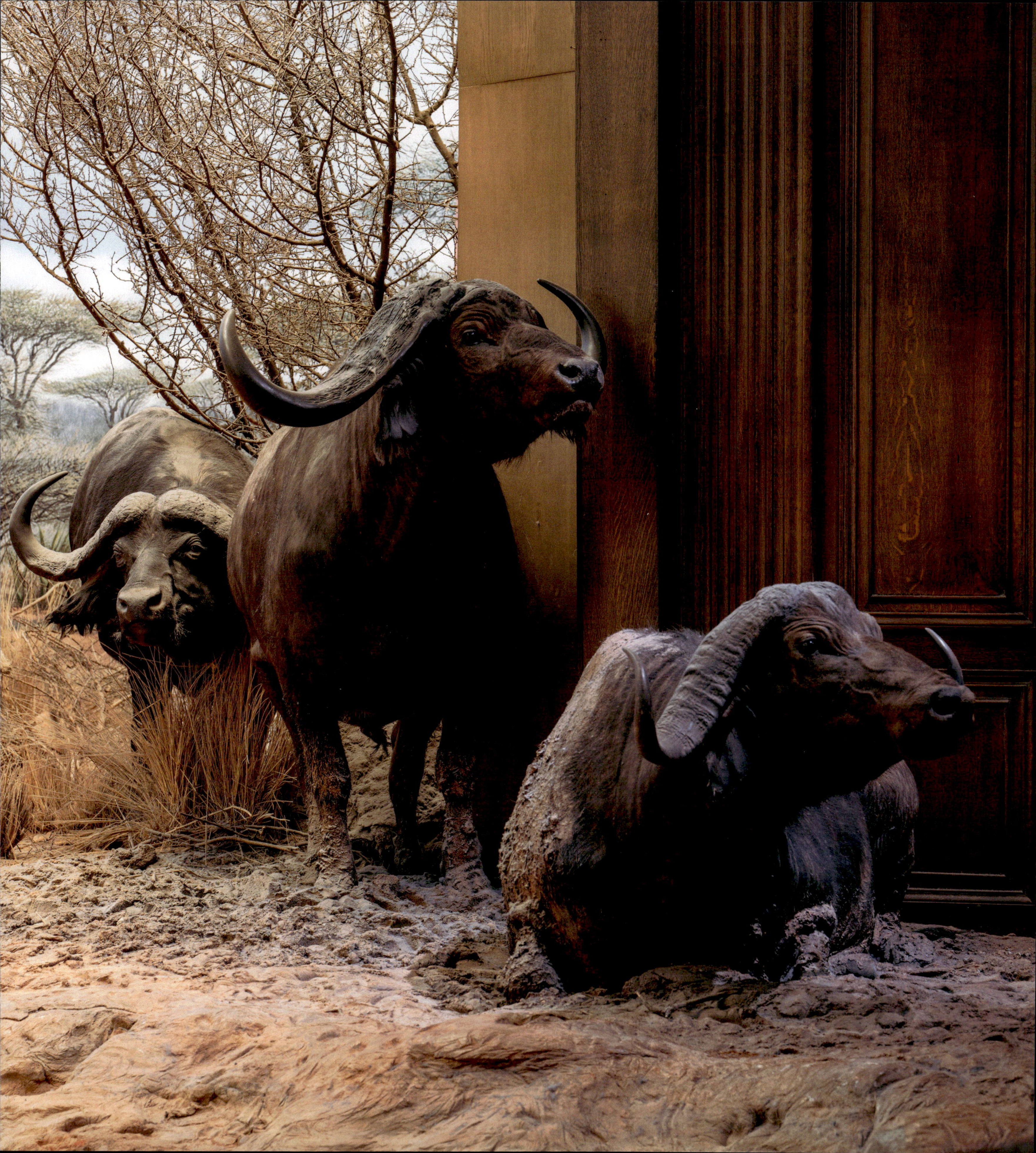

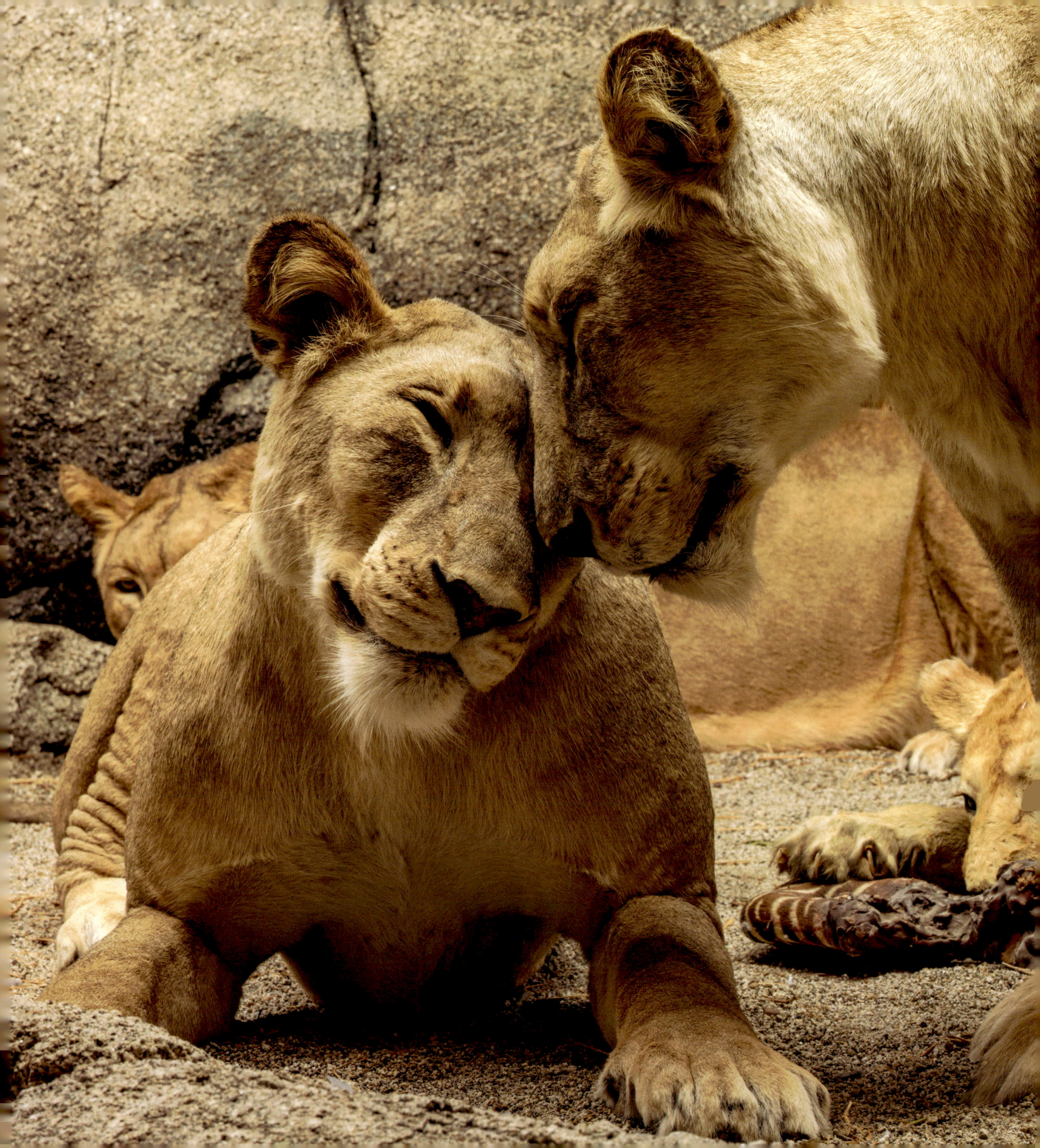

Habitat Change

Matt Davis

← Two females nuzzle each other in the *Lion* group. Added only in 2018, these lionesses exemplify NHM's long history of updating and creating new dioramas, a practice that extends into the present (2021).

Filling out a visitor survey in 2022, a museum patron wrote, "Mammal dioramas look the same as always, which was comforting." As heartwarming as this nostalgic observation was, it couldn't be further from the truth; most visitors don't realize it, but the museum's dioramas are updated and changed frequently. Although the diorama *halls* are 100 years old, most of the *dioramas* are much younger because the scenes change so often. Only two of the 17 dioramas in the Mr. and Mrs. Samuel Oschin and Family Hall of African Mammals date back to its opening as the Simson Hall of African Mammals in 1930, and even those groups, the *Greater Kudu* and *Sable Antelope*, have changed significantly over the last century.

NHM owes the malleability of its dioramas to a unique architectural quirk of its halls. Most dioramas in other museums are hermetically sealed behind large sheets of glass slanted out toward the viewer. This angle reduces distracting glare and reflections. Those dioramas are built once and then sealed up forever, time capsules of the exact moment of original construction. NHM's dioramas are different. Taxidermist John Rowley worried that the large panes of glass he wanted to use for the hall would crack under their own weight if tilted at an extreme angle, so he designed a system with large, vertical sash windows. Instead of the diorama being sealed up forever, artists could raise and lower the glass in front of each one whenever they pleased and easily change the scene inside.

Sometimes these changes are small and unnoticed, like the insertion of a tiny bird into the foreground foliage or the replacement of older taxidermy with newer, more lifelike mounts. Both the *Greater Kudu* and the *Sable Antelope* dioramas date back to the opening of the Oschin Hall of African Mammals nearly 100 years ago, but each group had some or all of its taxidermy replaced by more impressive mounts made decades later.

However, in many cases species or whole scenes are completely swapped in and out. The rhinos moved across the Oschin Hall to another bay with a new background and foreground. The polar bears moved from Level 1 upstairs to Level 2, and the chimpanzees and Arabian oryx moved from Level 2 downstairs to Level 1. The easy access afforded by the halls' unique sash windows likely led NHM to develop a much more fluid and iterative style of dioramas than other major museums. Many institutions understandably treat their dioramas as historical artifacts or sacrosanct works of art; they can be restored if their colors fade but never purposefully altered. All parts (the background, foreground, and taxidermy) were built at the same time, according to one artist's vision. NHM's groups are more like rough drafts that are edited and updated freely. They are collaborations among artists who lived decades apart. For some scenes, the foregrounds were built years after the original backgrounds were painted, or the animals were rearranged and recombined into new groups.

These frequent changes are partly due to NHM's dioramas starting out simpler than they appear today. Early artists did the best they could with the material at hand (foregrounds of cheap dried grass and a few large focal mammals) and hoped that future collections and increased budgets would allow improvements later. With so many empty bays to fill, they frequently had to resort to installing temporary and unfinished dioramas. Before the Oschin Hall of African Mammals was built, the museum installed previews of the African diorama scenes in what would later become the Level 2 North American Hall. These previews featured taxidermied animals with only minimal foregrounds and blank white backgrounds. Framed background studies painted in Africa by artist Maurice Logan gave visitors some idea of what the final scene might look like.

Once these dioramas were fully installed in their proper hall, they were still subject to revision. As the museum acquired better taxidermy mounts or background painters, scenes were updated and repainted. Accurate plant models replaced dried leaves, and the original large mammals were joined by supporting casts of birds, insects, and reptiles. Leftover mounts and foreground elements were recycled whenever possible. Once the Cape buffalo moved to a larger habitat in the Level 2 Exotic Mammal Hall, their original habitat in the Oschin Hall of African Mammals was turned over to three pygmy hippopotamuses. To convert the scene from the buffalos' Zambezi River on the southeast coast of Africa to the hippos' Lofa River on the northwest coast of Africa (2,000 miles [3200 km] and a hemisphere away), artists simply changed the location name on the diorama's label. When the *Becoming Los Angeles* history exhibition in the Kevin Sharer Hall needed a taxidermied ostrich in 2013 for a display about early amusements in the city, a male ostrich was borrowed from the Oschin Hall. One might never guess that the female brooding her eggs alone in the *Common Ostrich* diorama once had a mate.

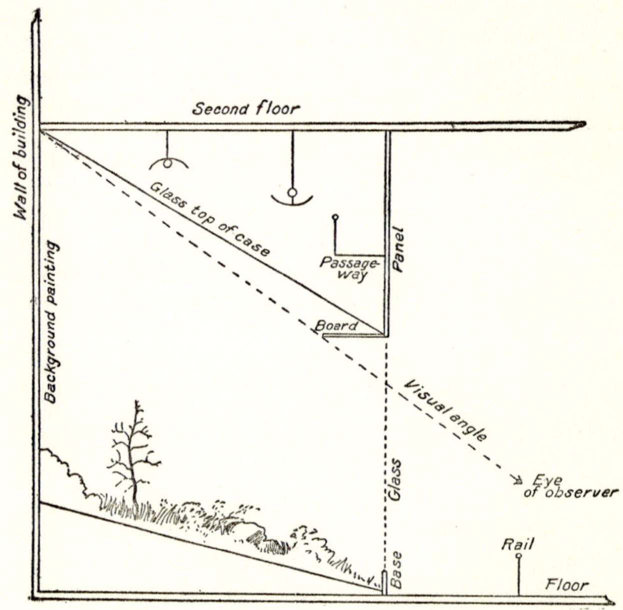

FIG. 17.—END VIEW OF MUSEUM GROUP.

↑ A diagram from John Rowley's 1925 book *Taxidermy and Museum Exhibition* shows his design for the NHM diorama bays, which are enclosed by vertical glass that can be raised and lowered like a window.

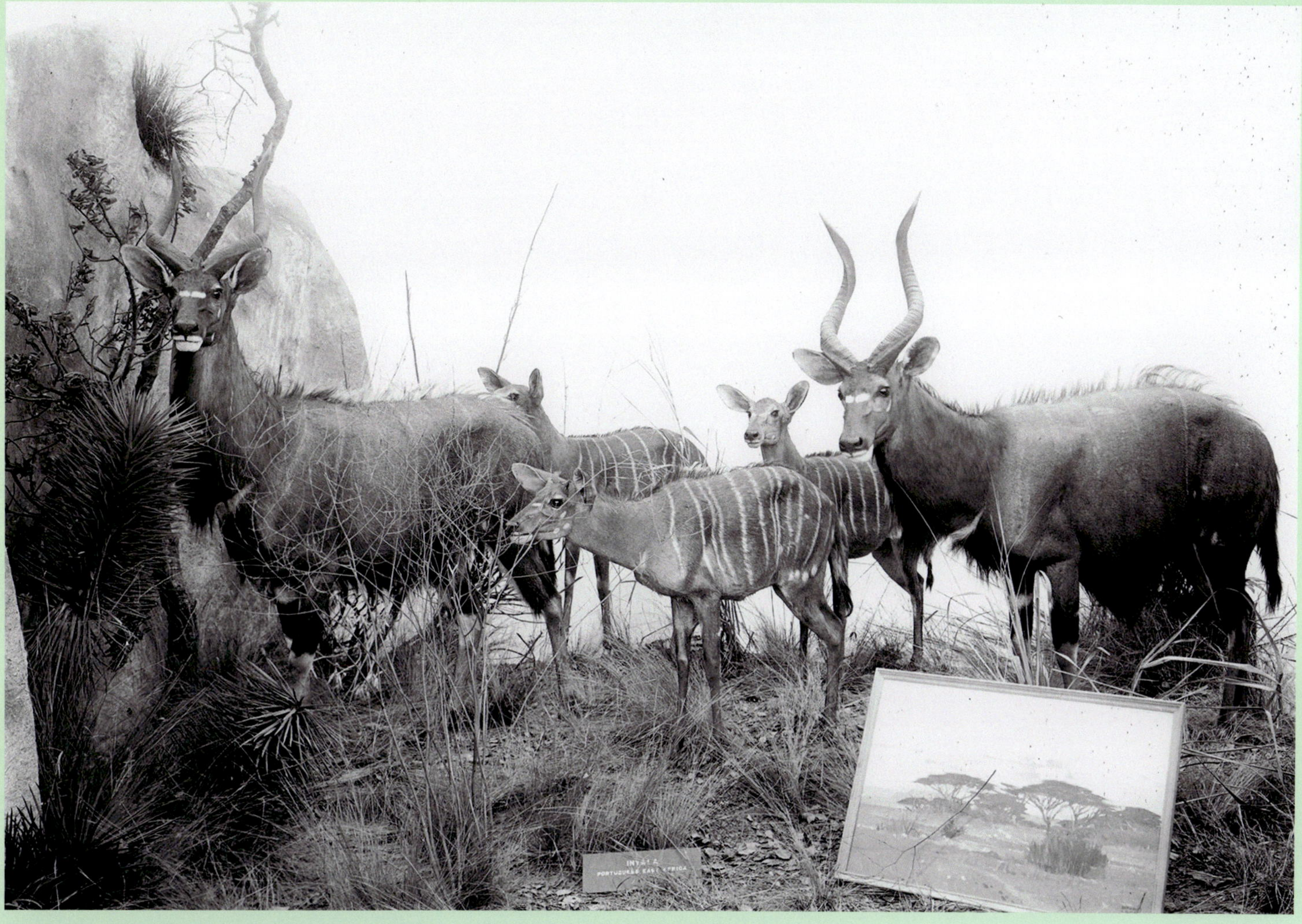

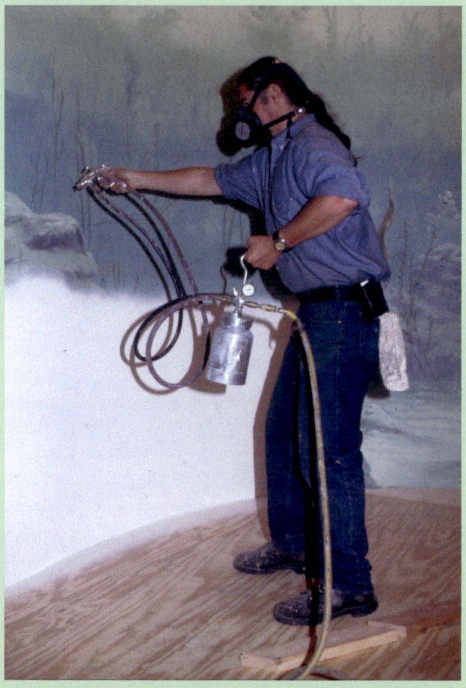

← The dioramas have never remained static; habitats were often completely replaced with new scenes. Artist Robert Reid covers an existing background with a white base before starting a new group (1990s).

↑ African groups like these nyala were originally installed in 1926 as preview dioramas with simple foregrounds, blank backgrounds, and study paintings showing visitors what the scene could eventually look like when complete.

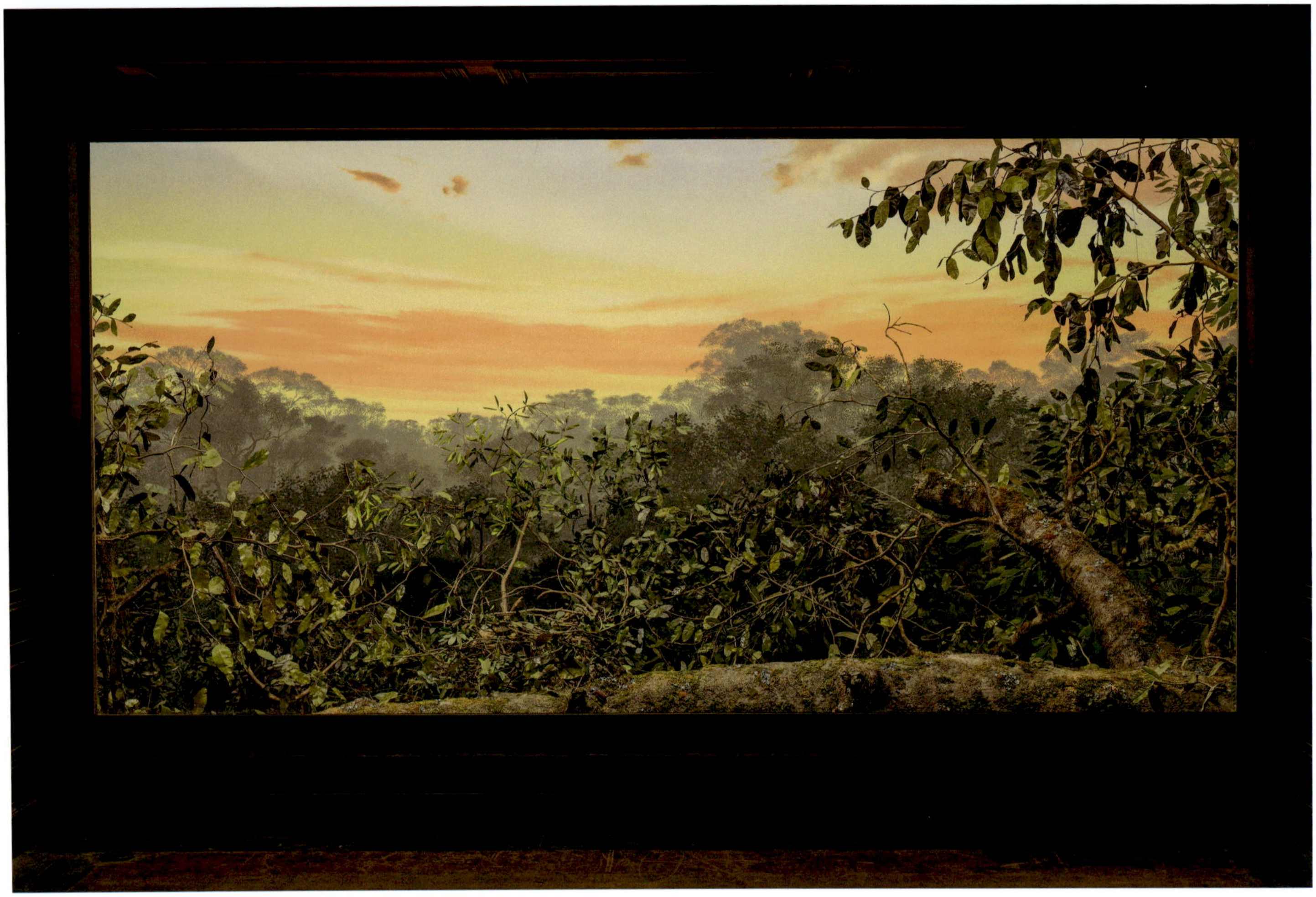

Looking back, all these changes may seem inexplicable. Why paint over an entire background? But NHM's diorama program is defined by artists who always strive to make the habitats more aesthetically pleasing, more scientifically accurate, and more diverse. The original halls were almost entirely the ungulates (hoofed mammals like deer or antelope) popular with big game hunters. But over time, artists dedicated groups to primates, rodents, and carnivores, filling out the mammal tree of life and the variety of habitats displayed. These changes break the mold of dioramas as still lifes and allow us to examine how the museum, and views on biodiversity, have changed over the century. Look closely the next time you are in the diorama halls; even though they appear old, there is probably something new that you haven't seen before.

↑ In 1992, the chimpanzees were moved from their original forest canopy setting in Uganda on Level 2 downstairs to a new Tanzanian habitat, leaving this empty diorama (2021).

↗ Hanson Puthuff's original 1930s background painting for the *Collared Peccary* group (c. 1955).

→ In the 1960s, an artist modified the *Collared Peccary* background, helping it tie into the foreground better by removing several creosote bushes and deepening the gully (2021).

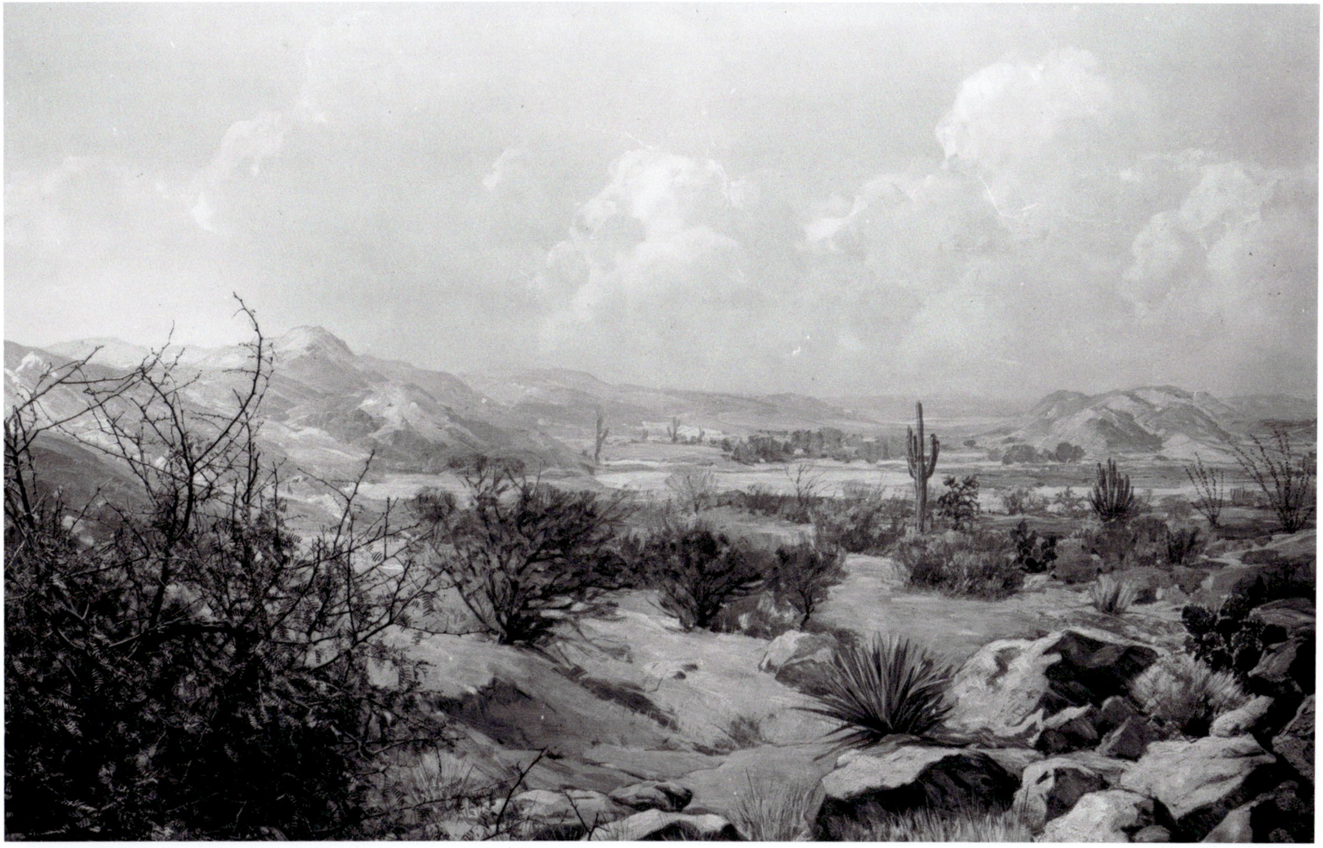

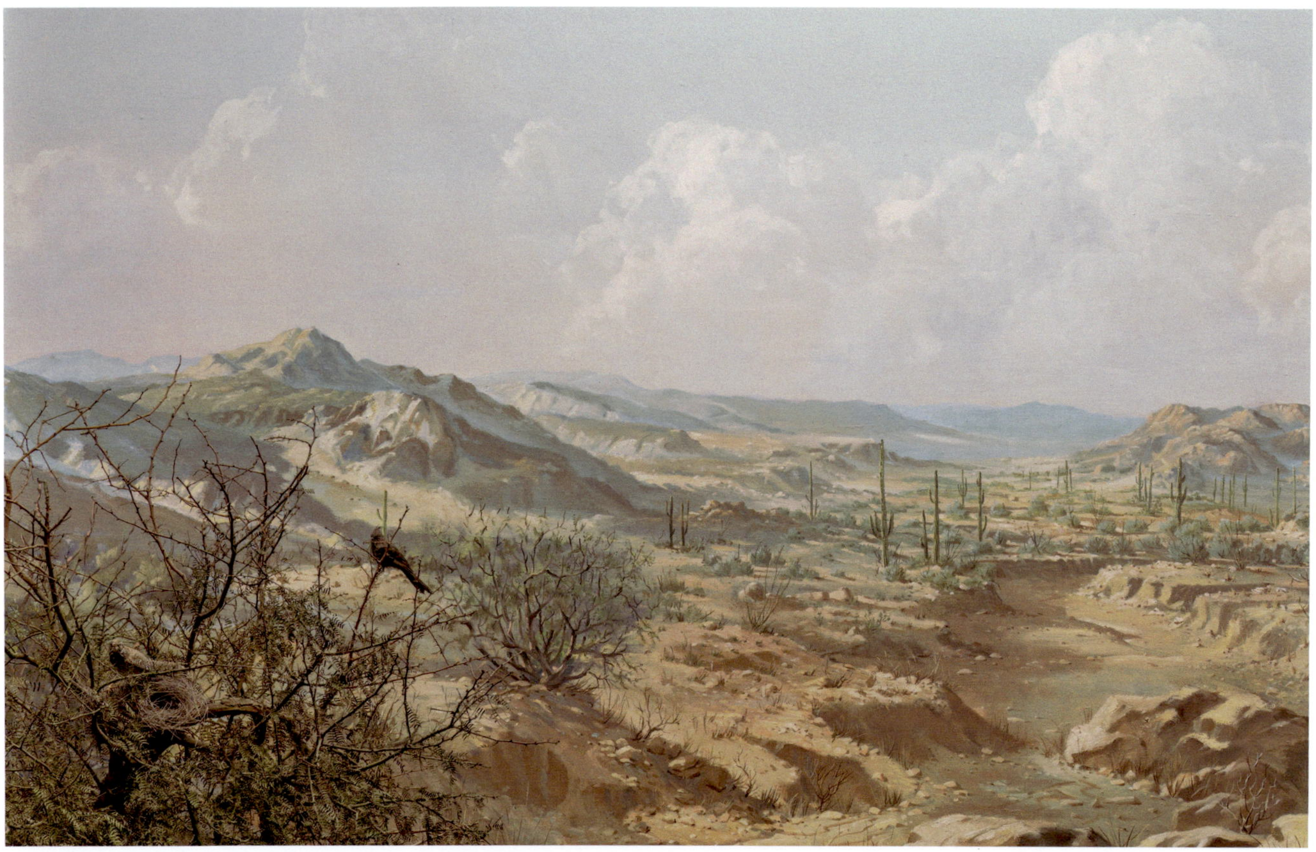

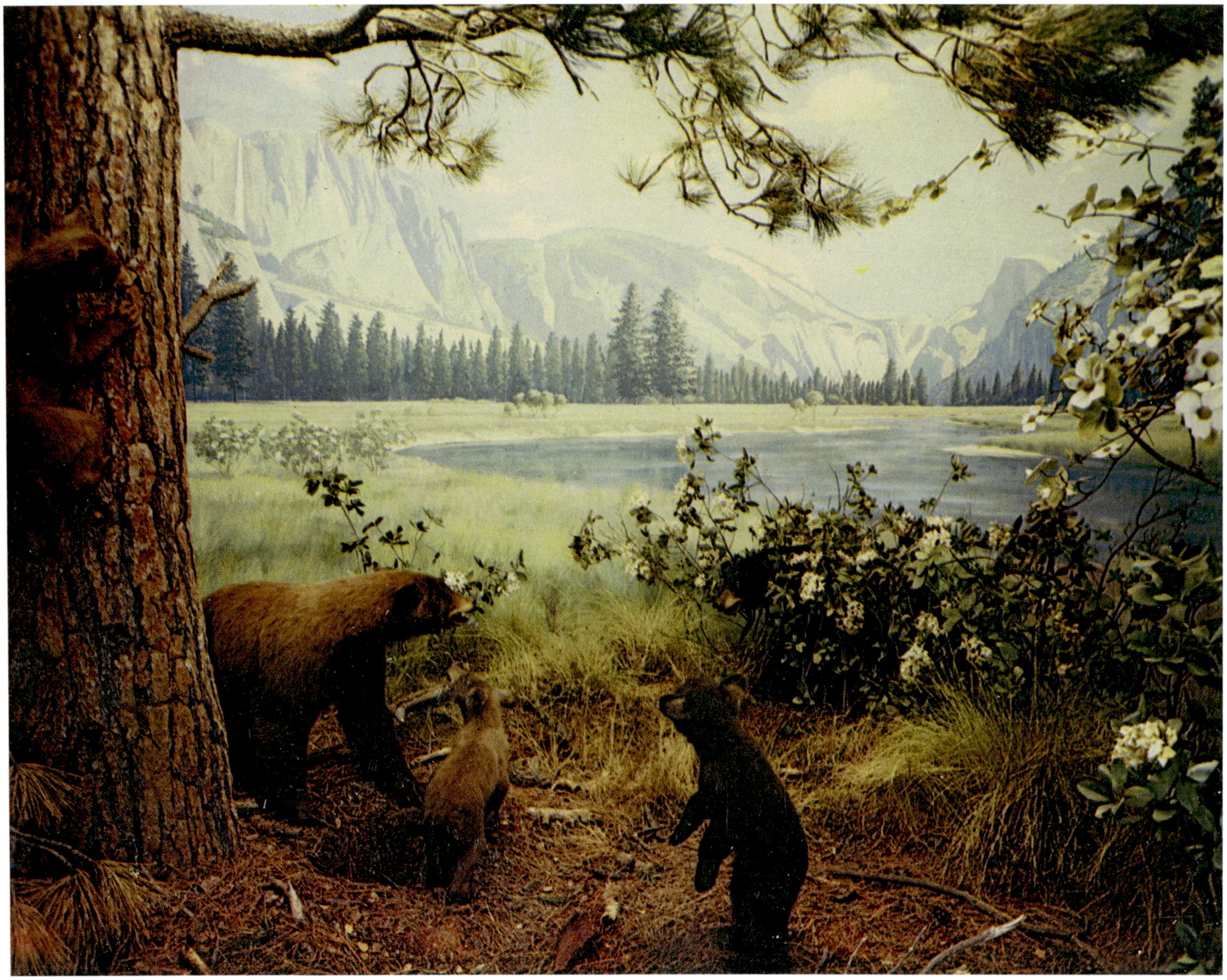

↑ The original 1951 *American Black Bear* group (1957).

→ In 1976, artists revised the *American Black Bear* group, keeping the Yosemite Valley scene largely the same but repainting the background and rebuilding the foreground to add in more detail (2021).

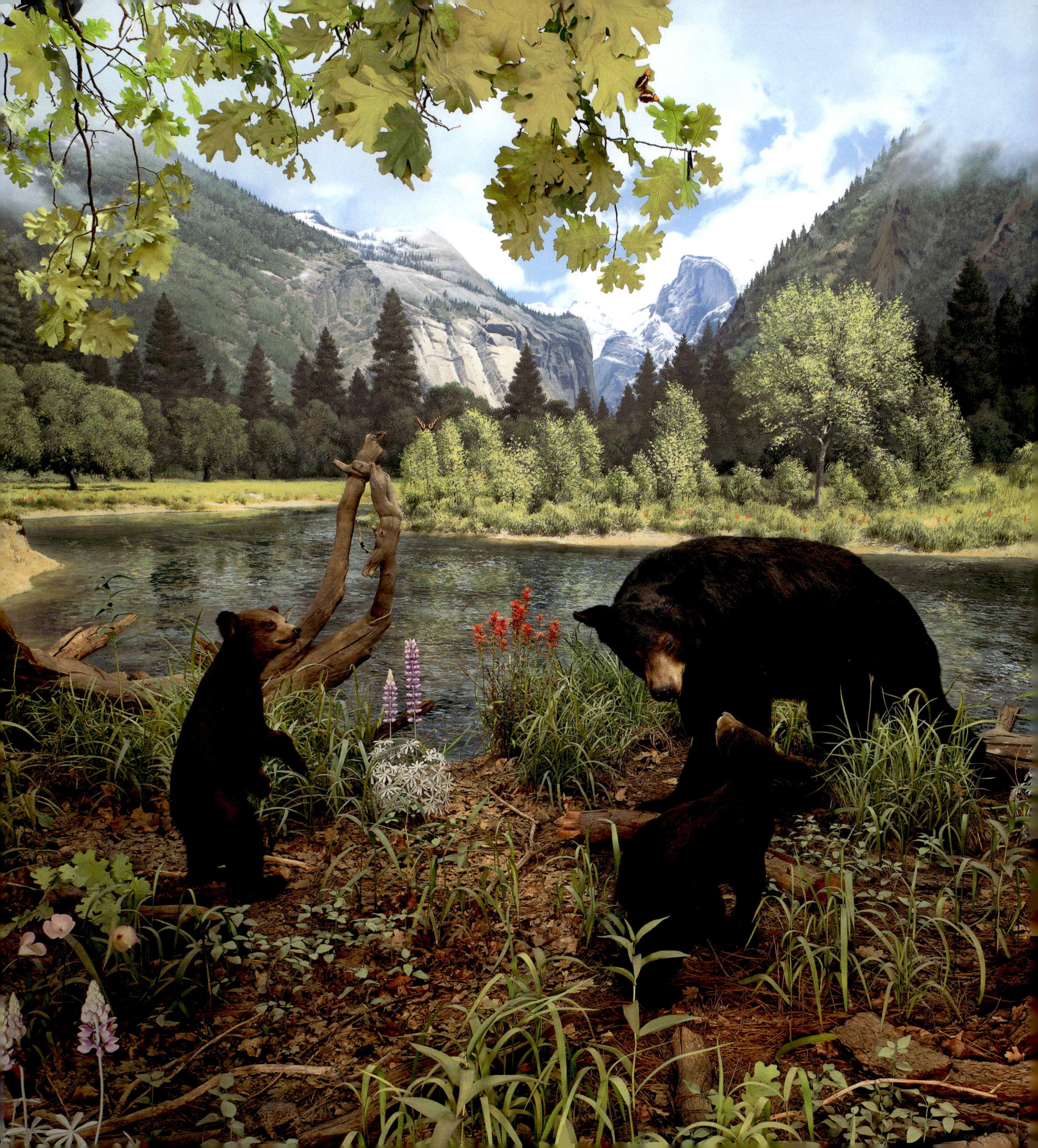

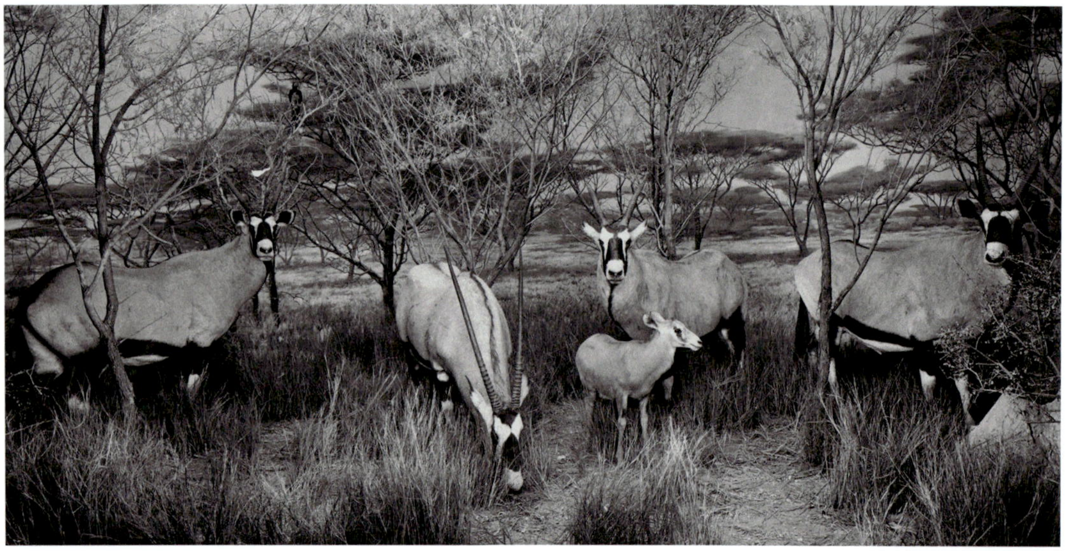

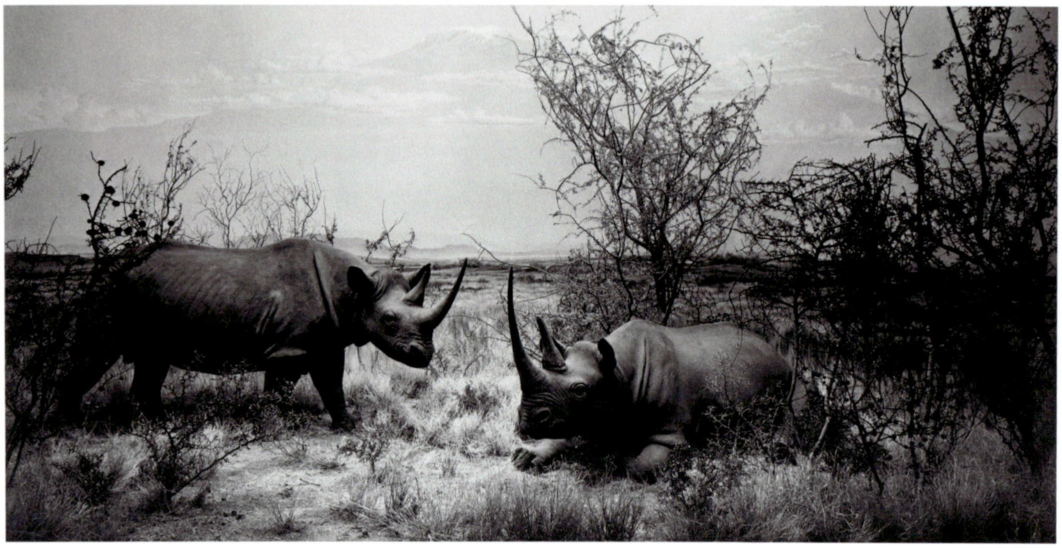

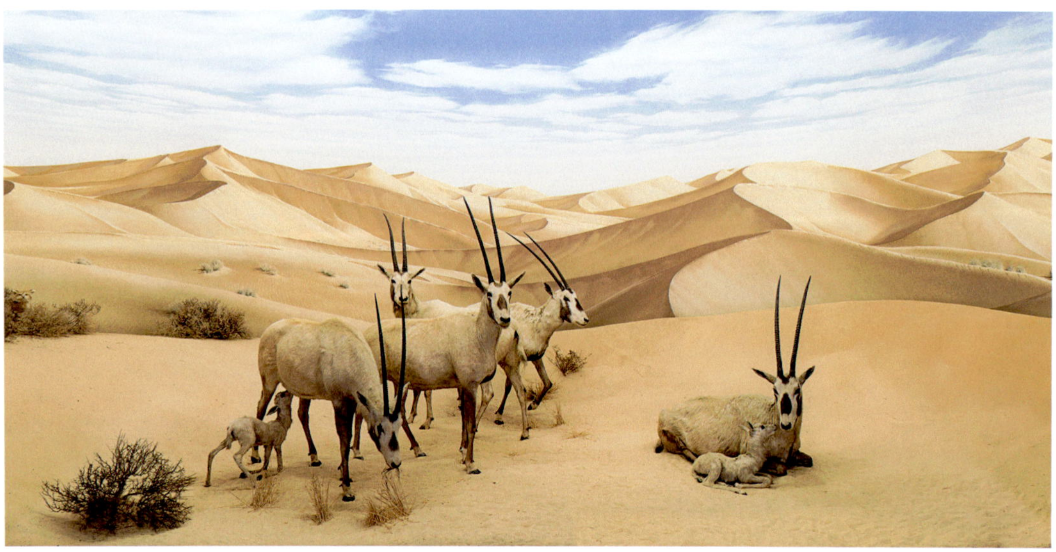

← The Oschin Hall of African Mammals has changed more than any other NHM diorama hall. The first diorama to the right of the entrance (Bay 1) has held three completely different habitat groups over the last century. The hall opened in 1930 with the *Gemsbok* group in Bay 1 (*top*, 1945), but it was moved upstairs to a new, more arid scene to make room for the *Black Rhinoceros* diorama, which opened in Bay 1 in 1958 (*middle*, 1965). The rhinos were then moved across the hall to Bay 11, where a nearly identical setting was constructed, so the *Arabian Oryx* group could appear in Bay 1 in 1988 as part of an overall rearrangement of the hall to make the habitats progress from wettest on the left to driest on the right (*bottom*, 2021).

→ Diorama revisions often moved taxidermied animals far from where they roamed while alive. The original *Pronghorn* group (*top*, 1930s or 1940s) opened in the Level 1 North American Mammal Hall and featured a desert scene from Modoc County, California, where the animals were collected. The animals were then moved up a floor in 1952 to a new diorama bay and a new sagebrush setting in nearby Lassen County, California (*middle*, 1950s). The herd was expanded with additional animals collected from Rawlins, Wyoming, in 1963, before the group was moved yet again to the San Francisco Peaks in Arizona in 1979 (*bottom*, 2014).

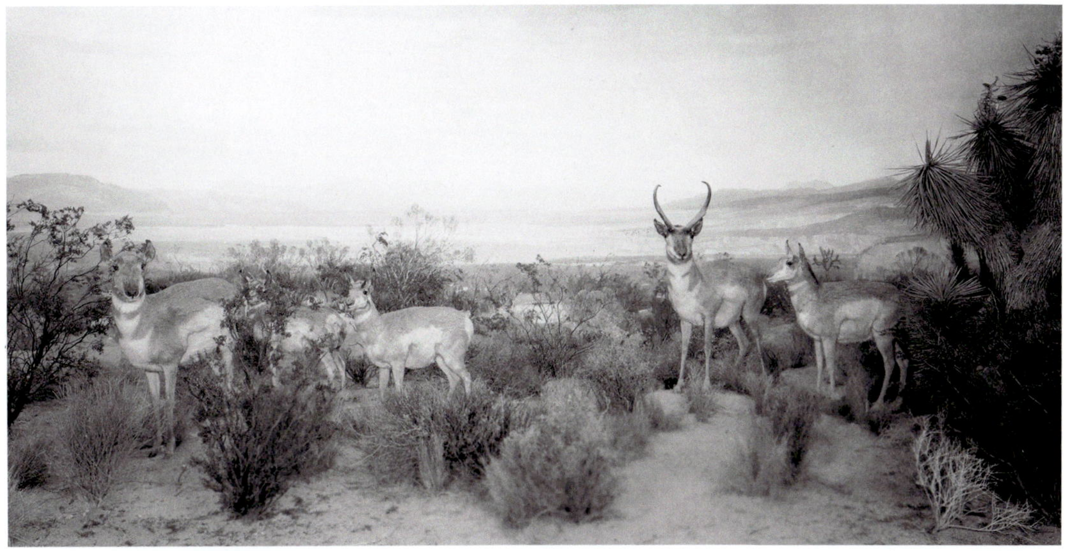

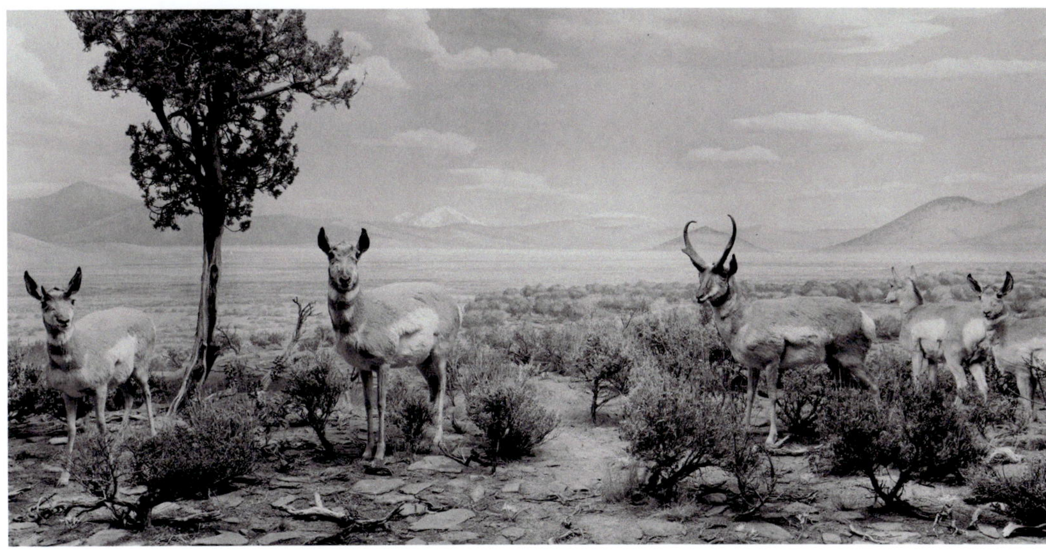

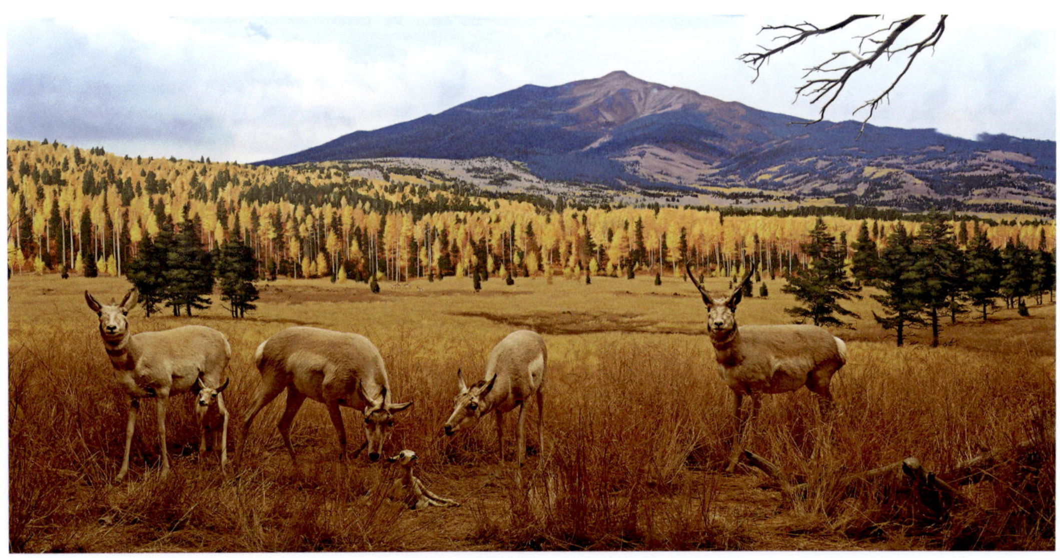

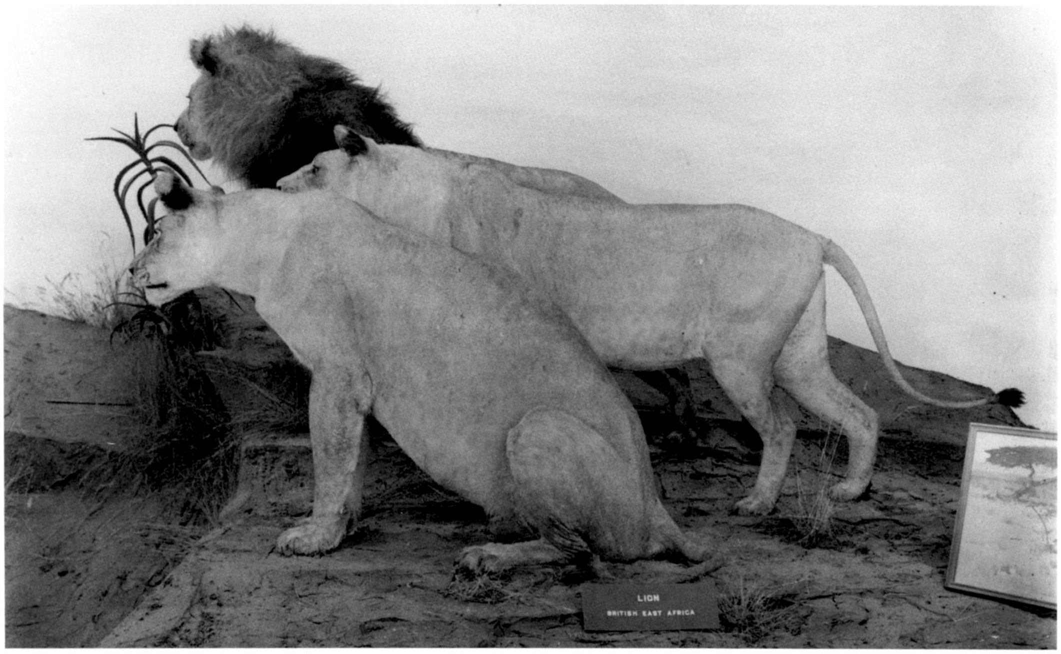

This spread
Probably no group has changed more than the *Lion* habitat. Although lions were collected for the original Simson Hall of African Mammals, they were installed only as a temporary preview group (*above*, 1926), and Bay 15 of the hall opened with a *White-tailed Gnu* group instead (*below left*, 1930). The scene used large trees to hide structural columns on both sides, but the forest habitat wasn't quite right for the gnu, a species more common in open plains. The gnu were replaced by a small family of lions and moved upstairs into a more accurate habitat (*below right*, c. 1940s). Unhappy with the existing male lion, though, NHM's artists taxidermied a more imposing mount in 1951 and tested it by hiding the lion in the bushes outside the museum. After the lion alarmed a few passersby and caused a leashed beagle to go absolutely berserk, the artists were confident they had created a lion scary enough, and swapped it into the scene (*facing page, top left*, 1973). But even he and his older mate were replaced in 1973 with more lifelike mounts, while rocks were added on the right to better cover the structural column on that side (*facing page, top right*, 1973). In 1989, the *Lion* group was moved directly across the hall to Bay 3 and completely redone, this time focusing more on scientific accuracy than beagle-scaring shock value. More lionesses were added to the scene to better reflect the actual social structure of lion prides, which are mostly females. The diorama was also opened up so that it spilled into the hall (*facing page, middle*, 1997). But open dioramas proved to be too much of a security concern, and the habitat was placed back behind glass in 2005. To more accurately reflect lions' behavior (they spend most of the day lounging around), two nuzzling females were added in 2018, with another male and lioness installed one year later. Rather than heroically roaring, the new male lies on his back, trying to scratch an itch. Today the scene appears authentic, although most of these animals never lived together in the wild. After all, the cubs were collected in the 1920s and are now 60 years older than some of their parents (*facing page, bottom*, 2021).

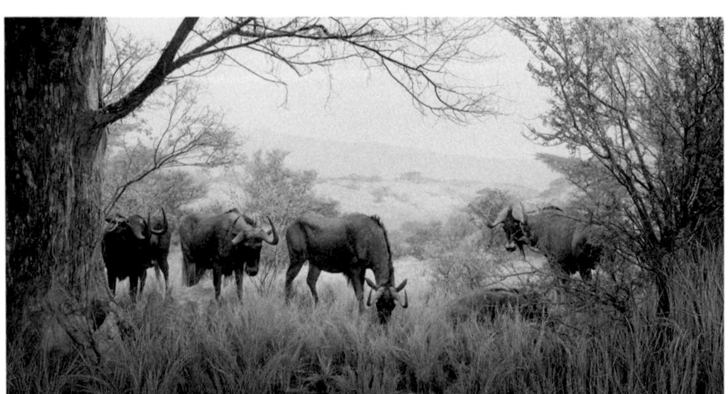

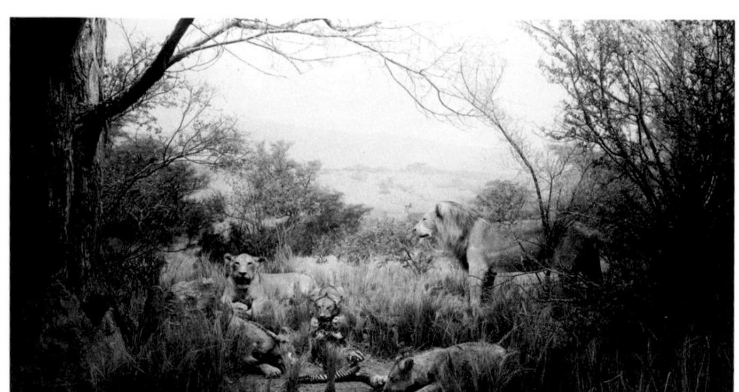

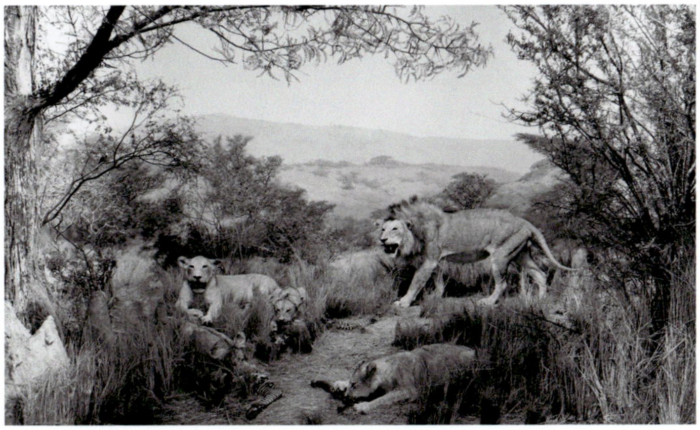
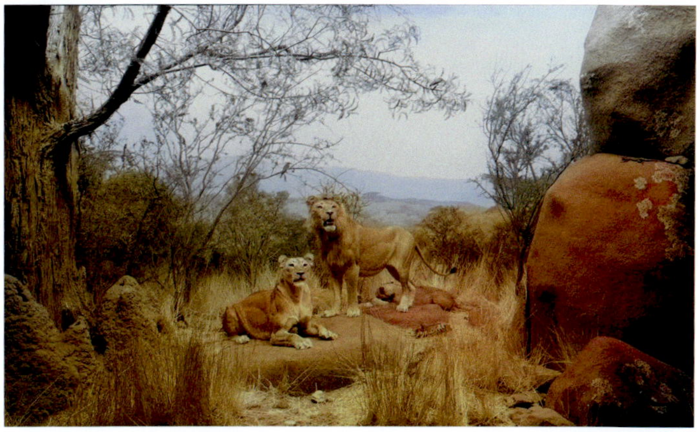
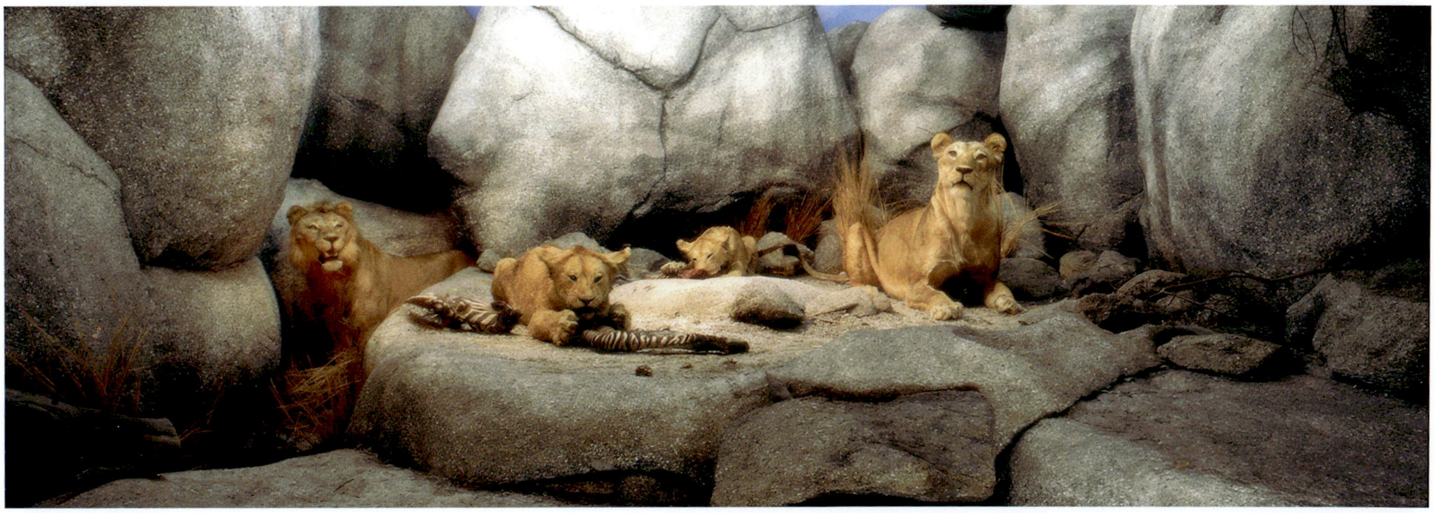
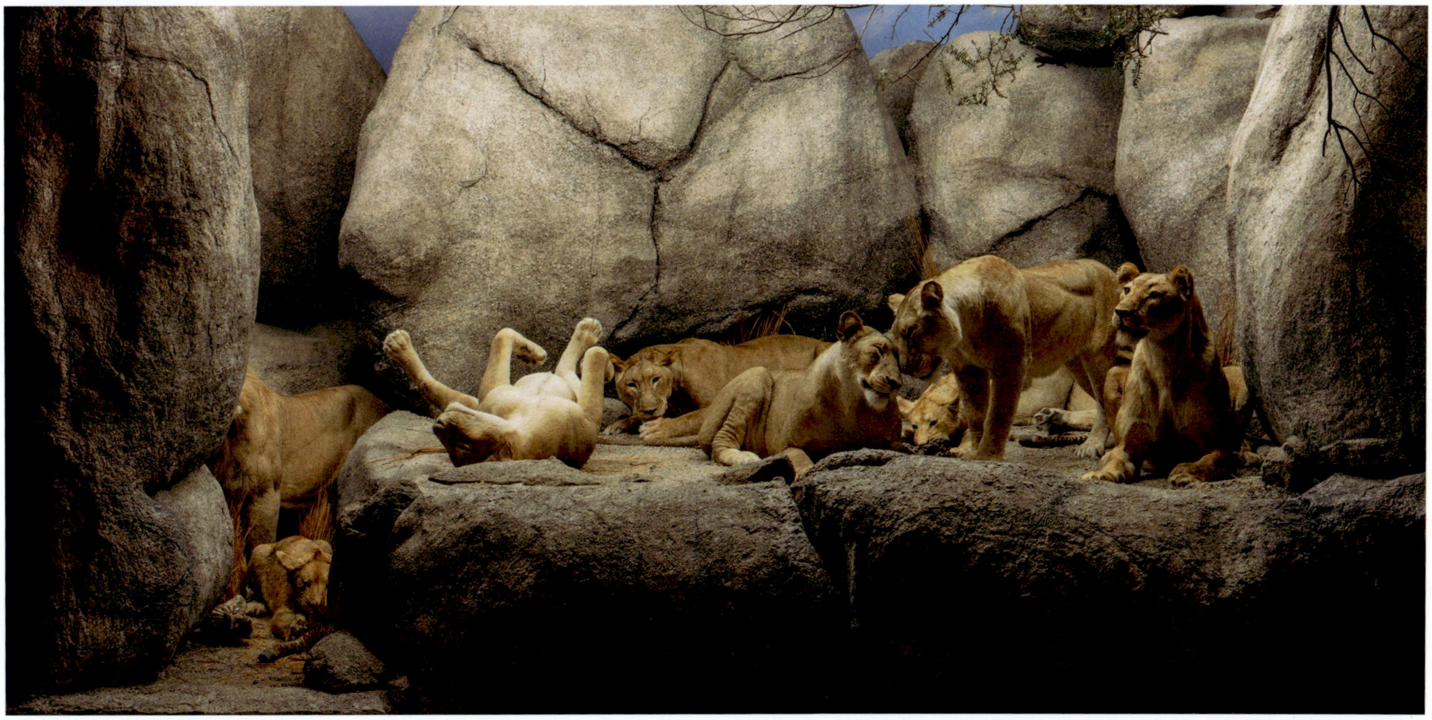

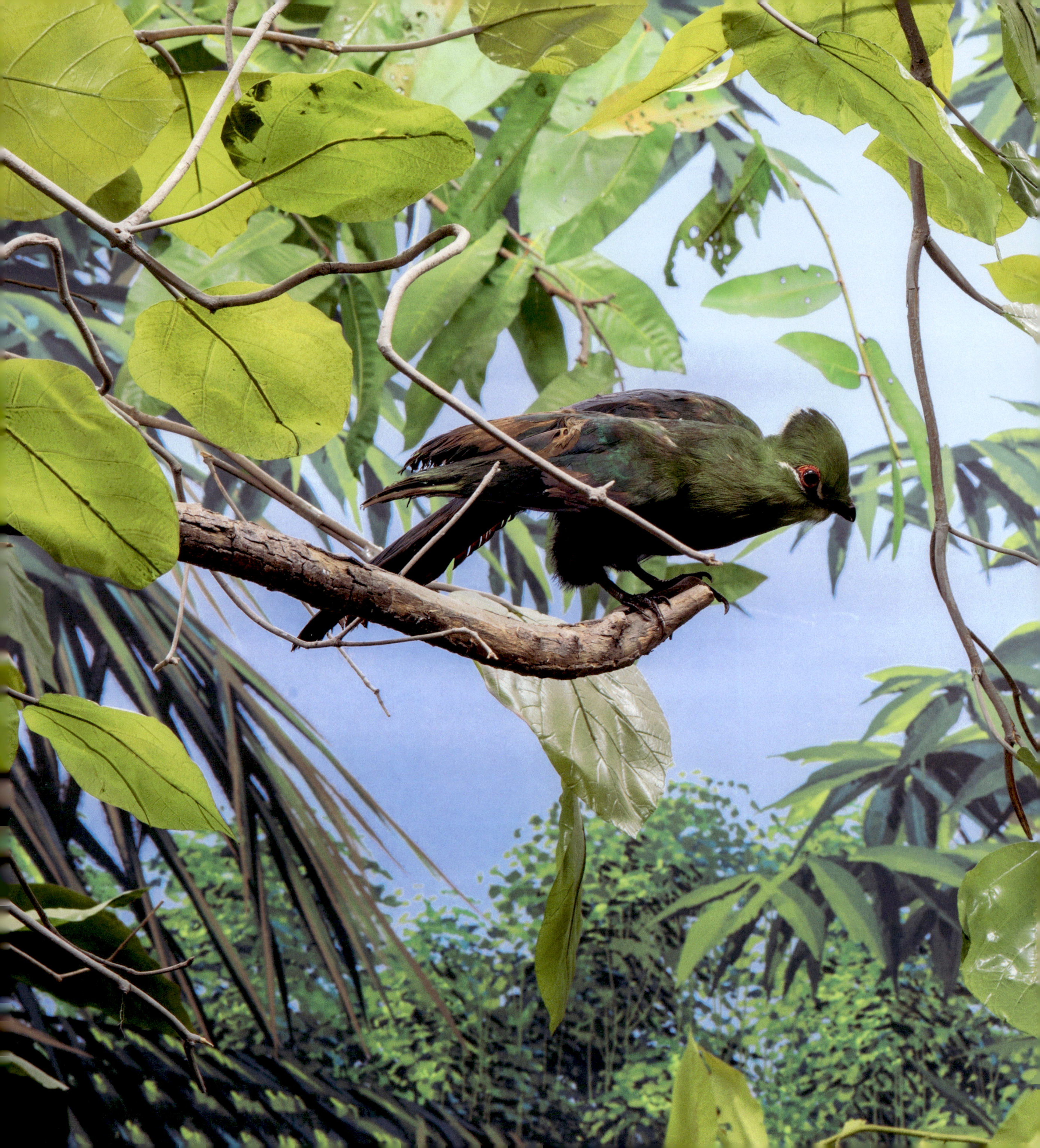

The Science behind the Scenes

Allison J. Shultz and Kayce C. Bell

← A black-billed turaco (*Tauraco schuettii*) perches expectantly in the *Chimpanzee* group. The large, charismatic mammals in dioramas were often used to excite the public and donors about expeditions to collect smaller animals, like birds, which are important for ongoing biodiversity research (2021).

Most people coming to NHM to see the Gem and Mineral Hall or the Jane G. Pisano Dinosaur Hall may not know that the exhibits are only one part of the museum. Behind the scenes are shelves and cabinets full of millions of specimens used for scientific research. There are rocks, fossils, pinned insects, jars of fish, and skeletons of everything from whales to hummingbirds. But when NHM first opened in 1913, the museum's collections were on display, and visitors would have seen cabinets full of all types of specimens. These early displays may have excited the educated specialist, but museum leaders like Taxidermist John Rowley thought long labels and rows of specimens bored the public because "they did not tell a story."[21] Aiming to engage all visitors with visual stories, he helped create the diorama halls, which were likely the first displays at NHM that did not appear like its research collections. Although they may look different, NHM's diorama halls are still intimately linked to the museum's research and the specimens that it has collected over the last century.

Specimens on display and specimens behind the scenes are quite different. Animals in dioramas are works of art that use taxidermy to create lifelike illusions; they are posed and given fake tongues and glass eyes. Specimens for research are often prepared as "study skins";

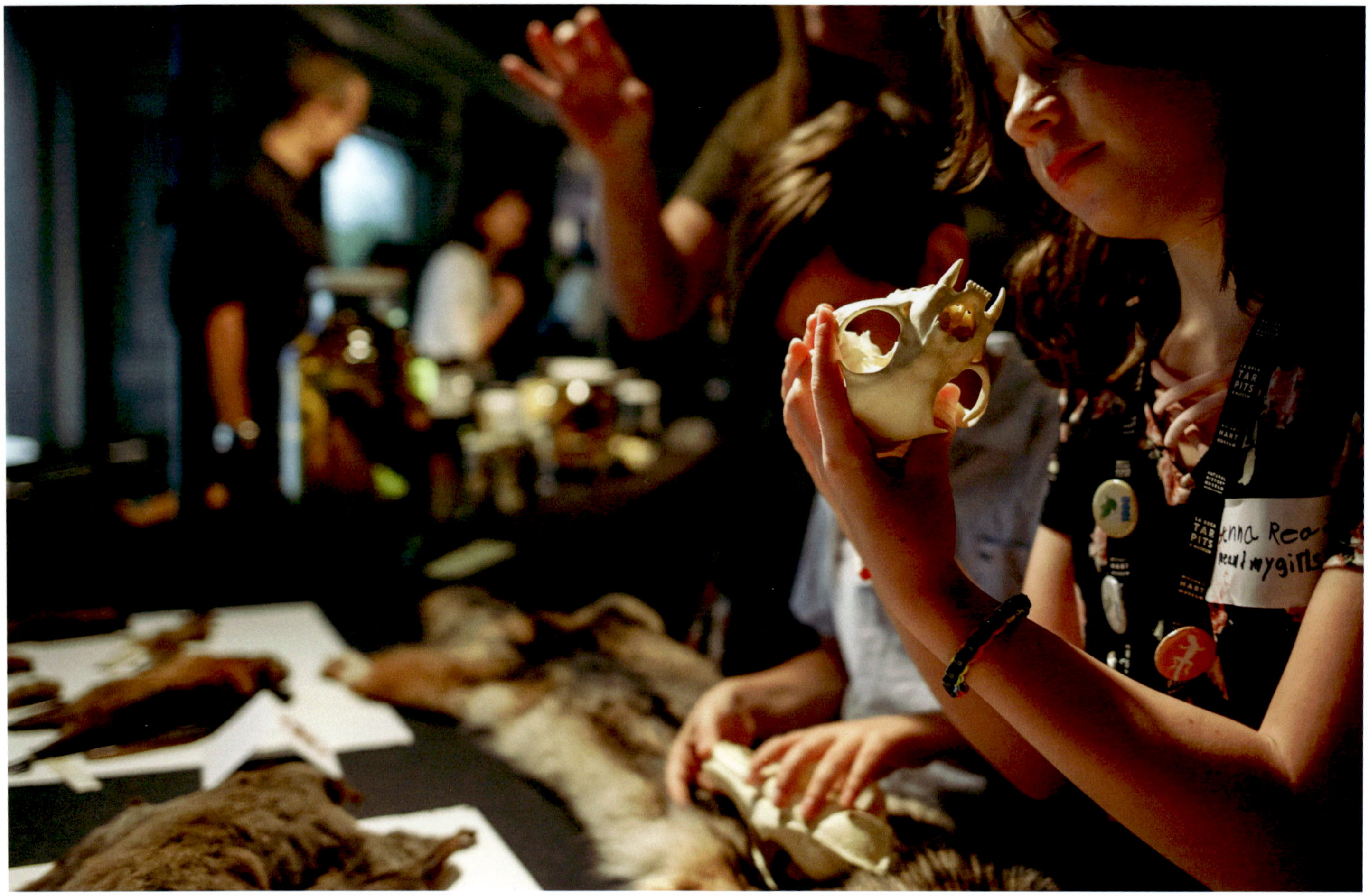

↑ A museum visitor closely examines a mammal skull and other behind-the-scenes specimens at a research collections open house held in the Level 2 Hall of North American Mammals (2019).

Following spread
Some of the bird and mammal species collected in the Itama region of the Bwindi Impenetrable Forest of Uganda on a 1967 expedition.

they don't need to appear lifelike to be useful to researchers. Bird and mammal study skins are stuffed with cotton in uniform shapes to maximize space in collection drawers while leaving most of their physical traits (ear shape, number of toes, wing length, etc.) accessible for measurement and analysis. Reptiles, amphibians, and fish usually have their whole bodies preserved in jars of alcohol; insects are pinned out and dried; and plants are dried and pressed onto paper. A single animal might even be preserved in various ways in multiple different collections. For example, the skins of mammals may be used for realistic taxidermy in dioramas, their skeletons stored in the Mammalogy research collections, and their parasites, like ticks and botflies, in the Entomology research collections.

Using a specimen for both display and study is not the only way dioramas are connected to the research collections. They were also crucial for driving many of the museum's earliest and largest research expeditions. They used large, charismatic mammals like tigers and giraffes to generate excitement with the public as well as to inspire donors to fund expeditions to collect small mammals, birds, and insects. These expeditions surveyed the biodiversity of an area; some specimens became part of a diorama to create a realistic replica of the habitat, but most entered the research collections to preserve a scientific record of biodiversity for the future. For example, the 1966 Cheney Kenya Expedition was organized to obtain two leopards for a diorama. But scientists also collected over 150 species and subspecies of birds for the Ornithology research collections. These specimens, and others collected on such expeditions, are critical for a wide variety of research uses, including a verifiable record of biodiversity at a specific place and time. They can even lead to the discovery of new species, like the

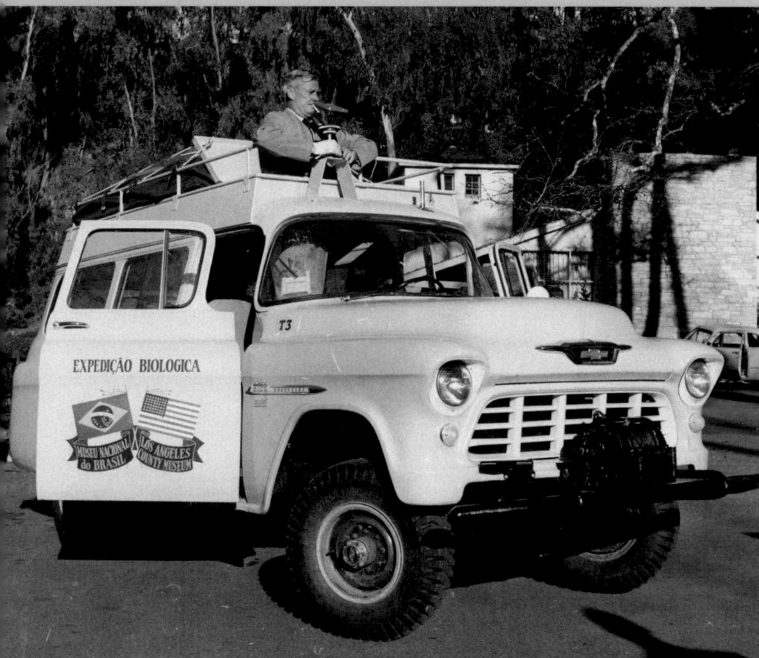

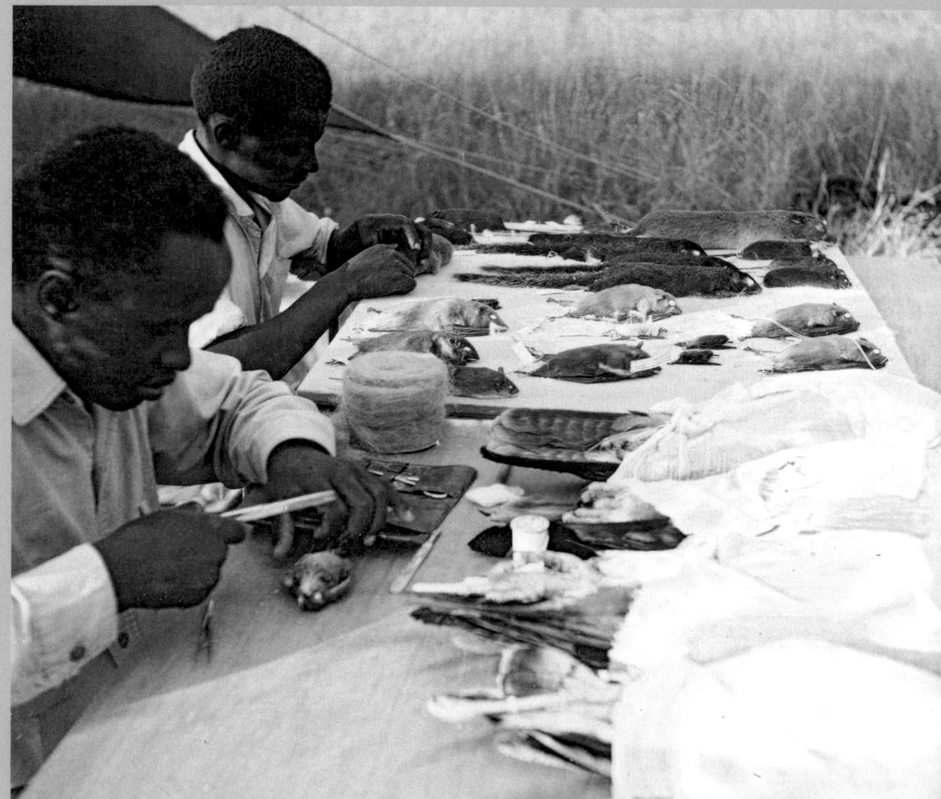

↑ Maurice Machris tests one of the four Chevrolet trucks custom built for the 1956 Machris Brazilian Expedition, a joint effort between NHM and the Museu Nacional do Brasil.

↗ NHM's expeditions often enlisted the aid of local institutions like the Coryndon Museum of Nairobi, Kenya, which had its skilled staff prepare small mammals and birds in the field on the 1964 Cheney Tanganyikan Expedition.

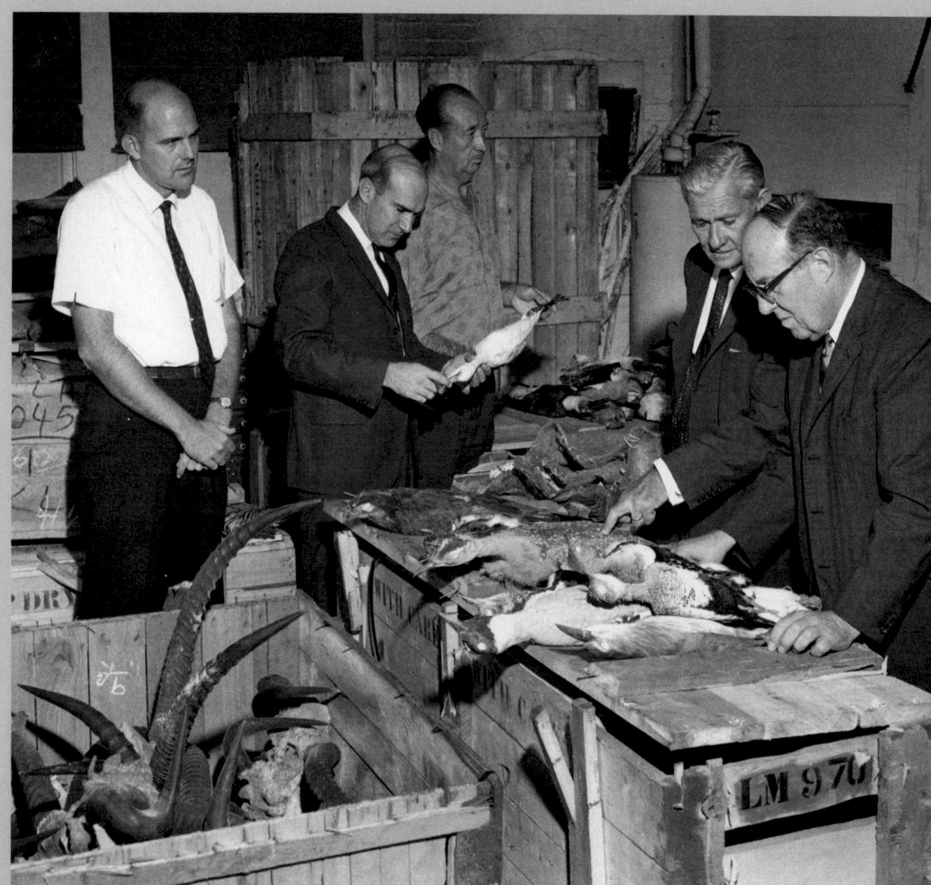

→ Machris (*second from right*) and Director Herbert Friedmann (*right*) examine bird specimens just shipped back to the museum from the 1963 Knudsen-Machris East Africa Expedition.

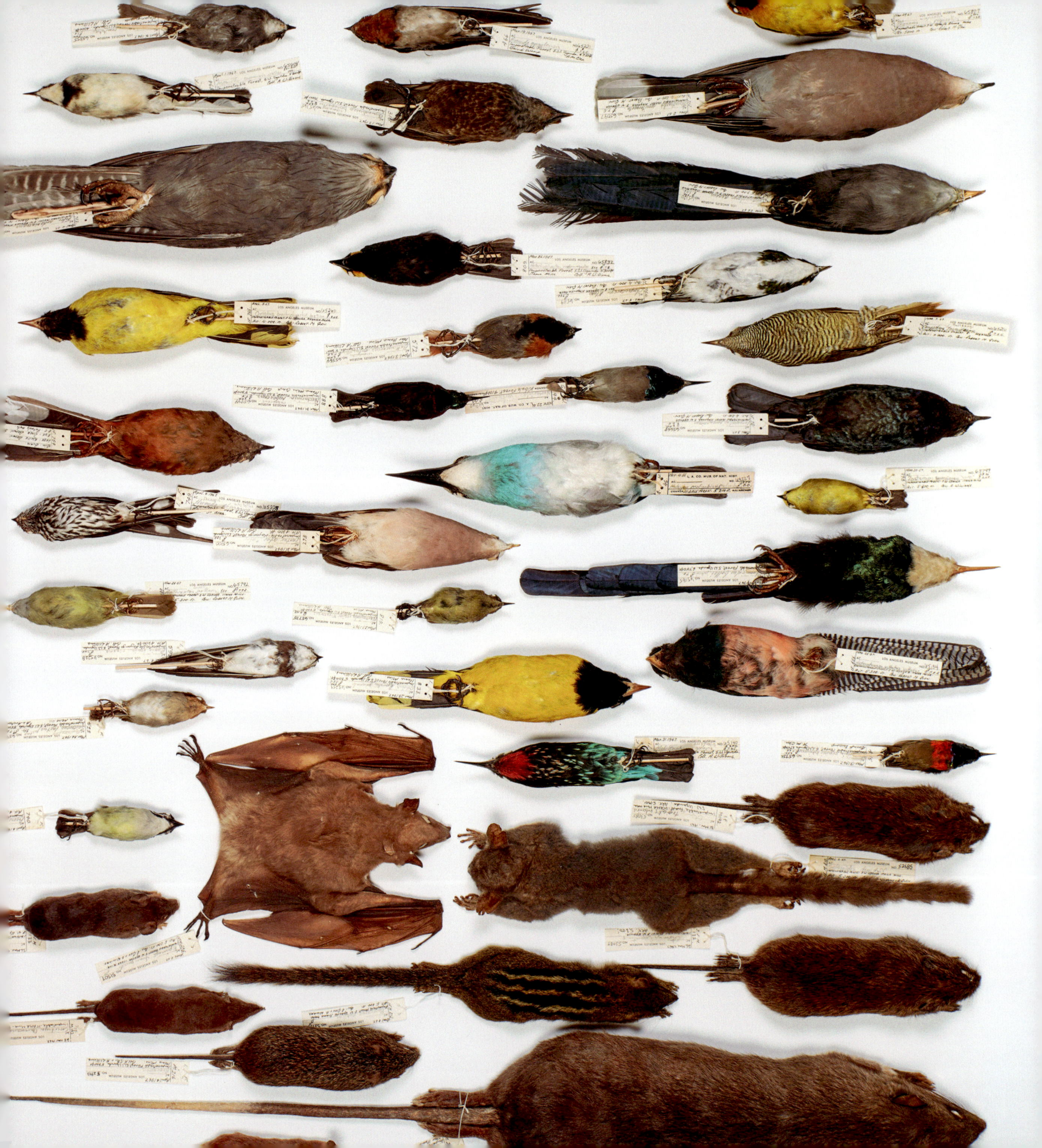

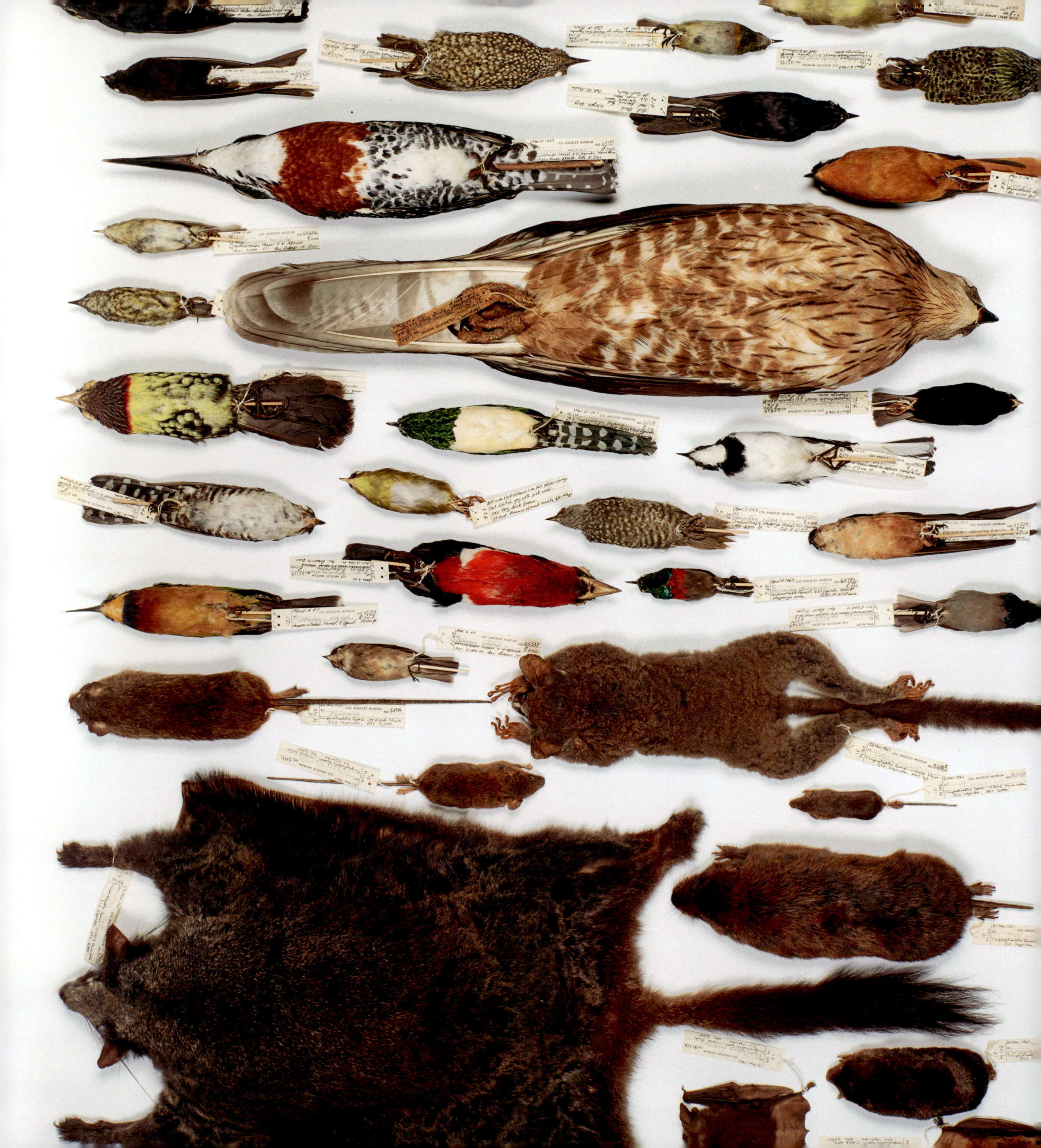
LOS ANGELES MUSEUM

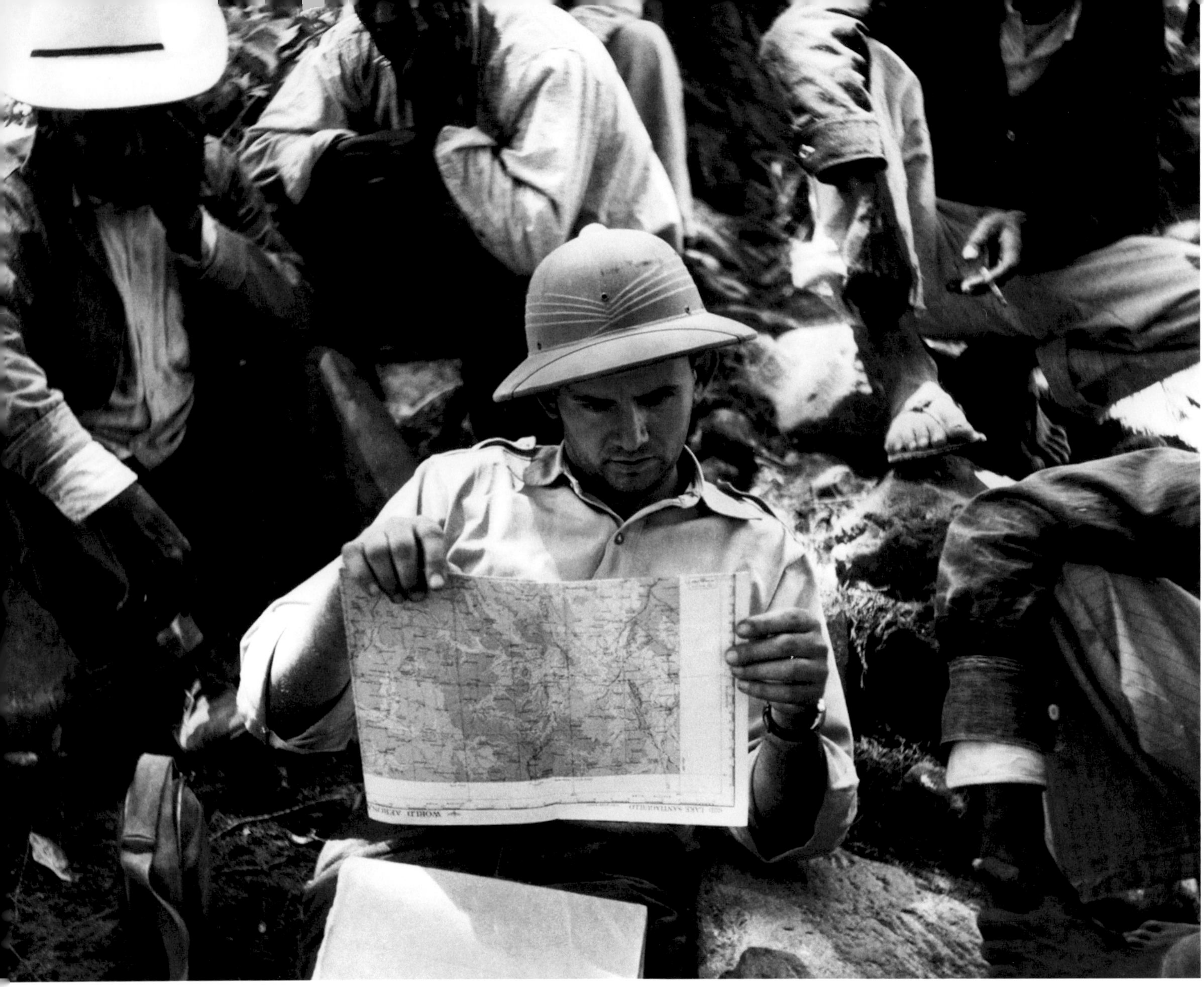

common kusimanse (*Crossarchus obscurus*), a type of dwarf mongoose, which was described from a museum specimen collected on an expedition to Uganda in 1968.

As tangible biodiversity records, specimens allow researchers to understand how organisms are responding to a changing planet. A century ago, species like the cactus wren (*Campylorhynchus brunneicapillus*) were common across the L.A. basin. Today, cactus wrens are restricted to small areas like the Palos Verdes Peninsula because their former natural habitat has been taken over by houses and roads. Scientists only know that the wrens used to be widespread because of 100-year-old specimens in the collections from places like San Fernando, Pasadena, and Lankershim (now North Hollywood). All specimens housed in natural history collections play an

↑ Curator of Birds and Mammals Kenneth Stager consults a map while his field team looks on during the 1950 Sheffler Barranca de Cobre Expedition in Mexico. An intrepid explorer, Stager led many expeditions for the museum, even managing to collect 139 bird specimens in Asia during World War II, where he served in US Army special ops units such as Merrill's Marauders. He infamously collected a silver pheasant (*Lophura nycthemera*) deep in enemy territory in what was then Burma, dutifully noting that it was "prepared during Japanese artillery barrage!" on the specimen tag (LACM Birds 20351, February 4, 1945; photo from 1950).

invaluable role in documenting life on Earth. They are like time machines that allow scientists to revisit past conditions, to investigate historical biodiversity and compare it to what we see today. This is becoming increasingly critical as scientists document and try to slow down or reverse the impact humans have on wildlife. The more specimens in collections, the more researchers can learn from them. For example, researchers can investigate how a bird like the house finch (*Haemorhous mexicanus*) has changed in Los Angeles over 100 years by looking at changes in size, beak shape, or feather color. With specimens from across its whole range, scientists can also understand how these traits vary around North America across sites that have experienced different kinds of environmental change. They can also compare it to the other species it interacts with or combine information about plants and animals within entire communities, just as the dioramas do.

High-quality data makes all of this research possible. It is essential to have the exact date and location associated with each individual specimen. Originally handwritten on tags or labels, much of the data is now digitized and searchable online so anyone can study where species live and their habitats. For example, researchers investigating the 1980 eruption of Mount Saint Helens used NHM Mammalogy collections data to show that the species of mammals that live around the volcano after its massive eruption are the same that lived

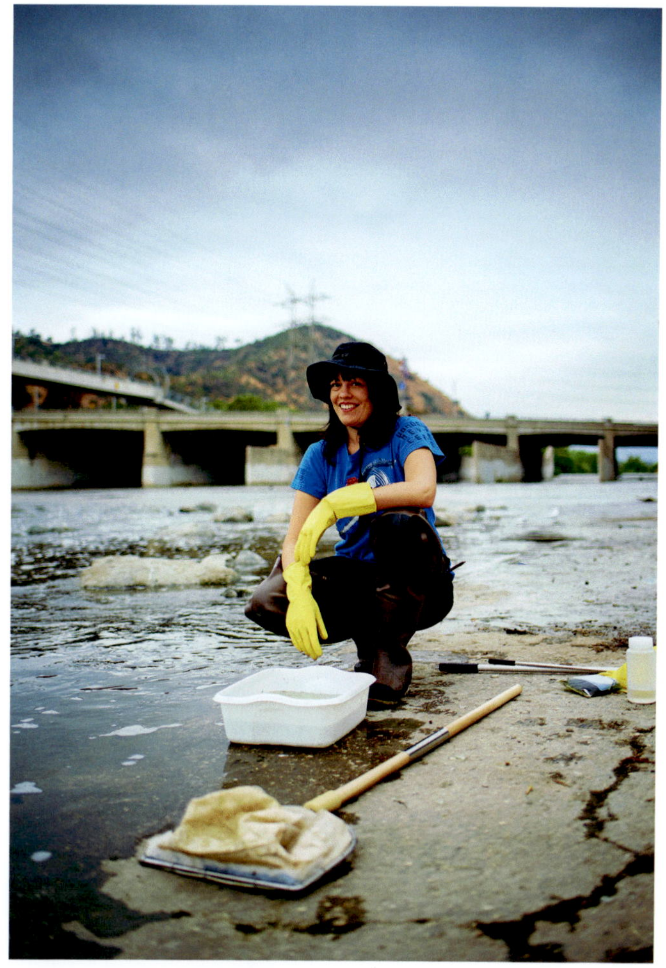

↑ Modern expeditions take place in both natural and urban areas. Here, NHM researcher Lisa Gonzalez collects insect samples from the L.A. River (2015).

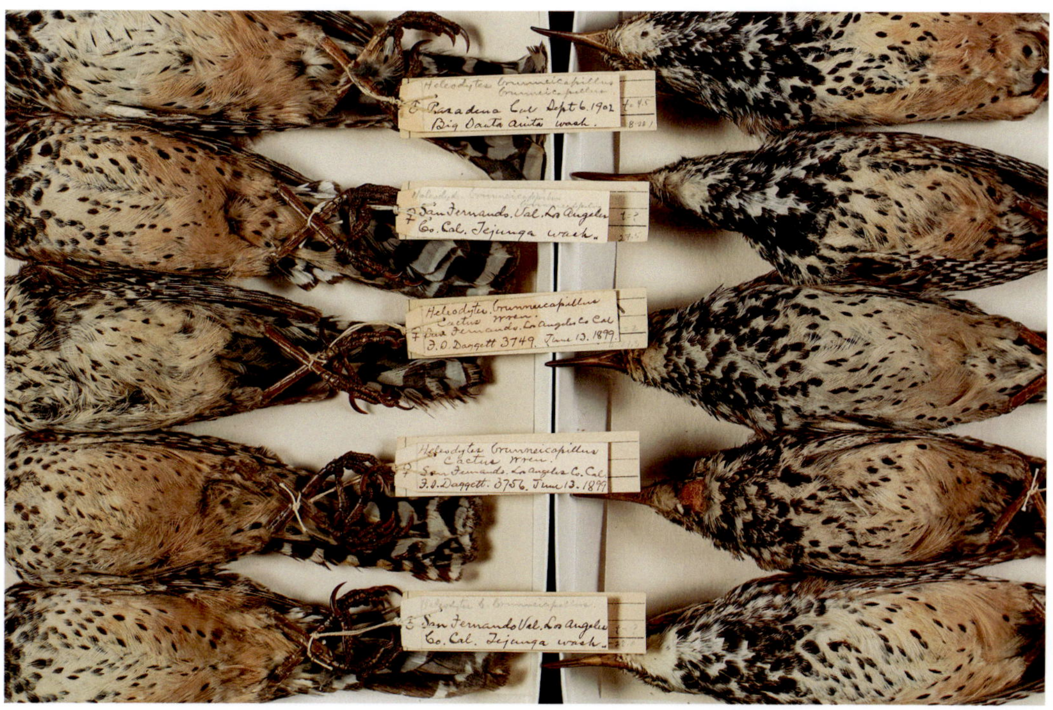

← When the museum's first director, Frank Daggett, collected these specimens in Pasadena and San Fernando around 1900, the cactus wren was common throughout the region. But it no longer is, due to habitat destruction. Specimens like these provide important records of historical biodiversity and remind us that environments are in flux (2023).

there before. Studies such as this use collections data to reveal how communities are able to recover from major events that substantially alter their habitat.

Scientists also use specimen data and records to identify which habitats to prioritize for conservation. In 2020, researchers "rediscovered" the Somali sengi (*Elephantulus revoilii*), a tiny mammal with a long nose, which was thought to have gone extinct fifty years before. Besides NHM, only seven other museums in the entire world had this poorly known species in their collections. Yet scientists were still able to combine these records to predict where the Somali sengi could live. Luckily, the rocky desert habitat that it loves is already in protected areas, and the species has a good chance of surviving in the wild. Occurrence data can also be used to predict hot spots of diversity where many species occur together. NHM bat specimen records, many of which were collected on diorama expeditions, contributed to a study that found rugged terrain is an important factor in bat species diversity in Africa. These are just a few examples of the impacts museum specimens and their data can have on understanding the natural world.

Just as the real specimens in dioramas impact visitors more strongly than two-dimensional images do, much more can be learned about a species from a physical specimen than from pictures. Traditionally, specimens were collected for taxonomic purposes like identifying new species and understanding species relationships. However, new technologies and new questions have only made physical specimens even more valuable. NHM's early scientists would still recognize some of the tools used by researchers today, like scales for weighing eggs and calipers for measuring the length of fish. But many modern tools are beyond anything they could have imagined. Researchers can now use computerized tomography (CT) scans to see inside an animal without dissecting it and scanning electron microscopes to see the tiniest details of insects. Researchers can extract DNA from specimens to learn about past population sizes or the evolutionary relationships among different species. Chemical analysis of fur and feathers (even those sampled from the animals in dioramas) can reveal an animal's diet or whether it lived in an area with heavy pollution. For example, seabirds in the NHM Ornithology

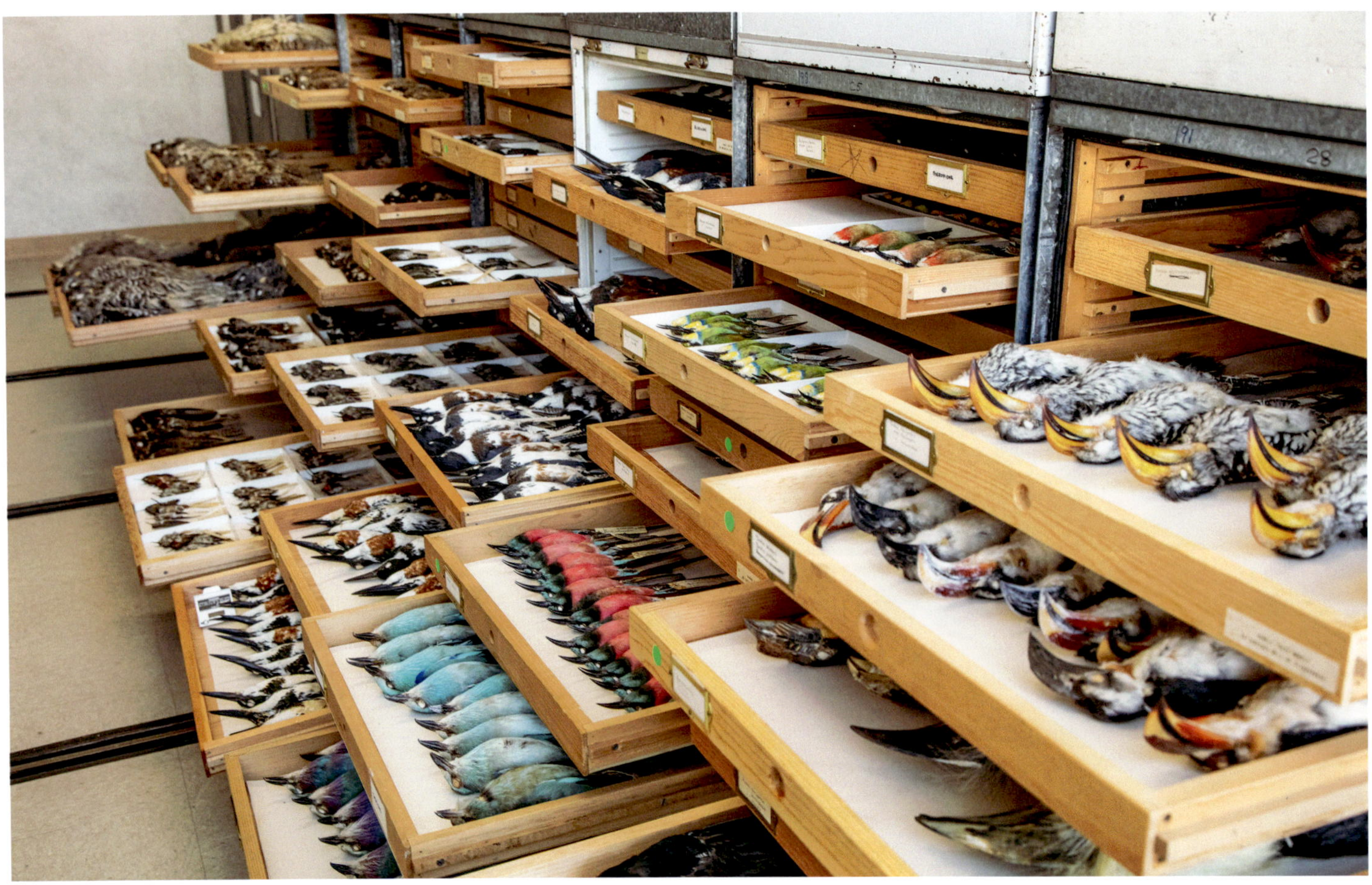

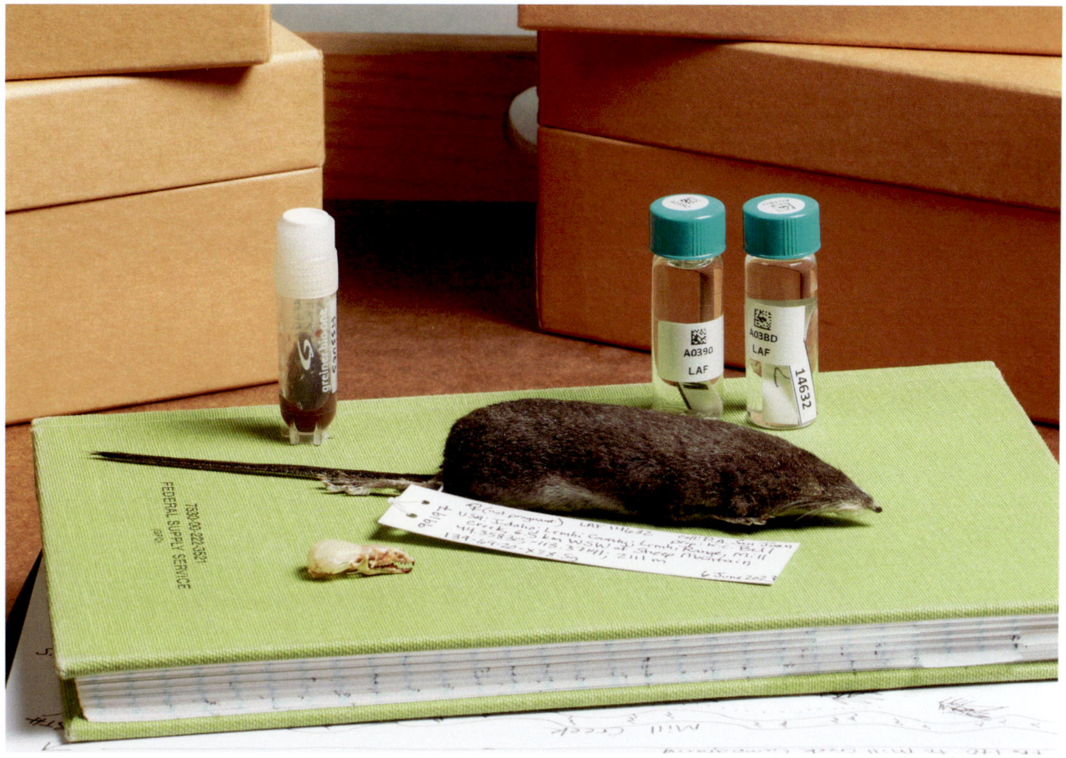

→ New technology and research questions have increased the parts of a specimen that are useful to preserve. Today, a typical specimen in the Mammalogy collections, like this American water shrew (*Sorex palustris*), might consist of a skeleton, a study skin, cryogenically frozen tissue samples, parasites like mites and tapeworms stored in alcohol, GPS coordinates, field notes, and 3D scans. All these data points can be combined in ways that collectors a century ago couldn't even imagine.

↙ Specimens in the Ornithology collections are time machines to specific places and dates. Each one, a unique representation of the environment and biology of that species, can be studied from scientific, historical, or cultural angles (2023).

collection have been used to show increases in mercury levels in the ocean over the last century. Far from becoming obsolete with technological improvements and digital data, physical specimens show scientists even more than they did a century ago. Besides skeletons and study skins, researchers now store high-quality tissue samples in cryogenic freezers so that they can be used later for sophisticated genetic analyses. They even preserve stomach contents to study animals' microbiomes.

Museum scientists and curators focus not only on scientific or societal needs of the present but also on those of the future. Like libraries designed to hold valuable knowledge forever, natural history collections preserve biological information in perpetuity. While the primary emphasis of the research collections is for scientific research, the collections are also used in other ways, such as for art and education. Visitors to the collections include artists working on paintings of birds, classes from local universities, and even Hollywood animators looking to create computer models of fur. Collections also have specimens of historical and cultural significance, like the pelt of the grizzly bear shot by a local Sunland farmer in 1916, which became the last confirmed grizzly in Southern California. The California grizzly bear (*Ursus arctos californicus*) was so important to California's identity that it was placed on the state flag, but that still wasn't important enough to save it from extermination. Specimens like the Sunland grizzly bear are tangible reminders of how quickly *our* local habitat can change. The biological archives held in collections serve as a repository for researchers to use now and into the future.

One past researcher who collected specimens for NHM was Joseph Grinnell of the University of California, Berkeley. In 1910, he predicted the following about the future value of specimens:

> This value will not, however, be realized until the lapse of many years, possibly a century, assuming that our material is safely preserved. And this is that the student of the future will have access to the original record of faunal conditions in California and the west, wherever we now work.[22]

Collections only increase in value over time, and just as the original dioramas were made to preserve habitats that were rapidly changing, it is important for museums to continue collecting specimens to understand how biodiversity is still changing. We owe it to the "students of the future."

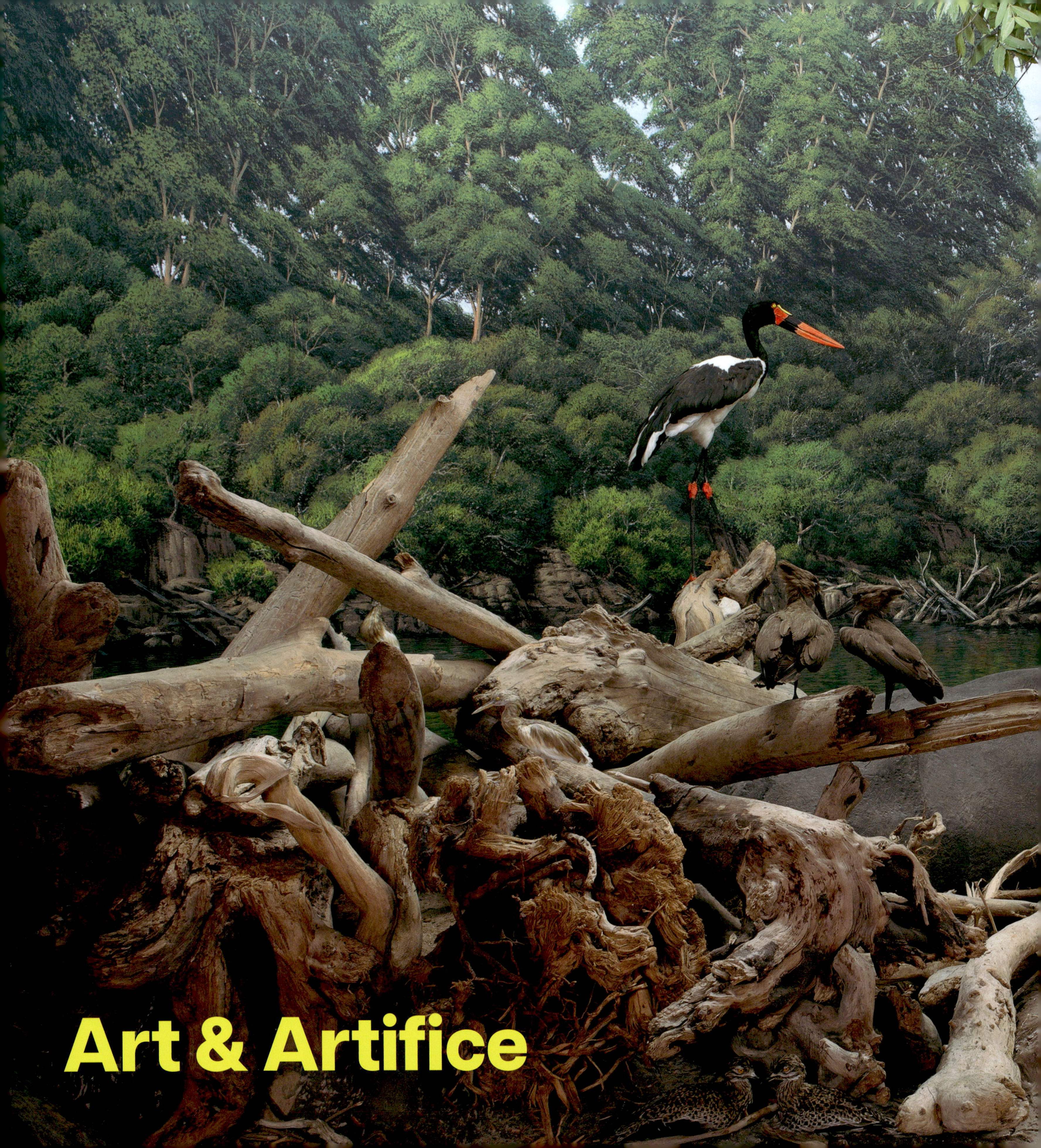

Art & Artifice

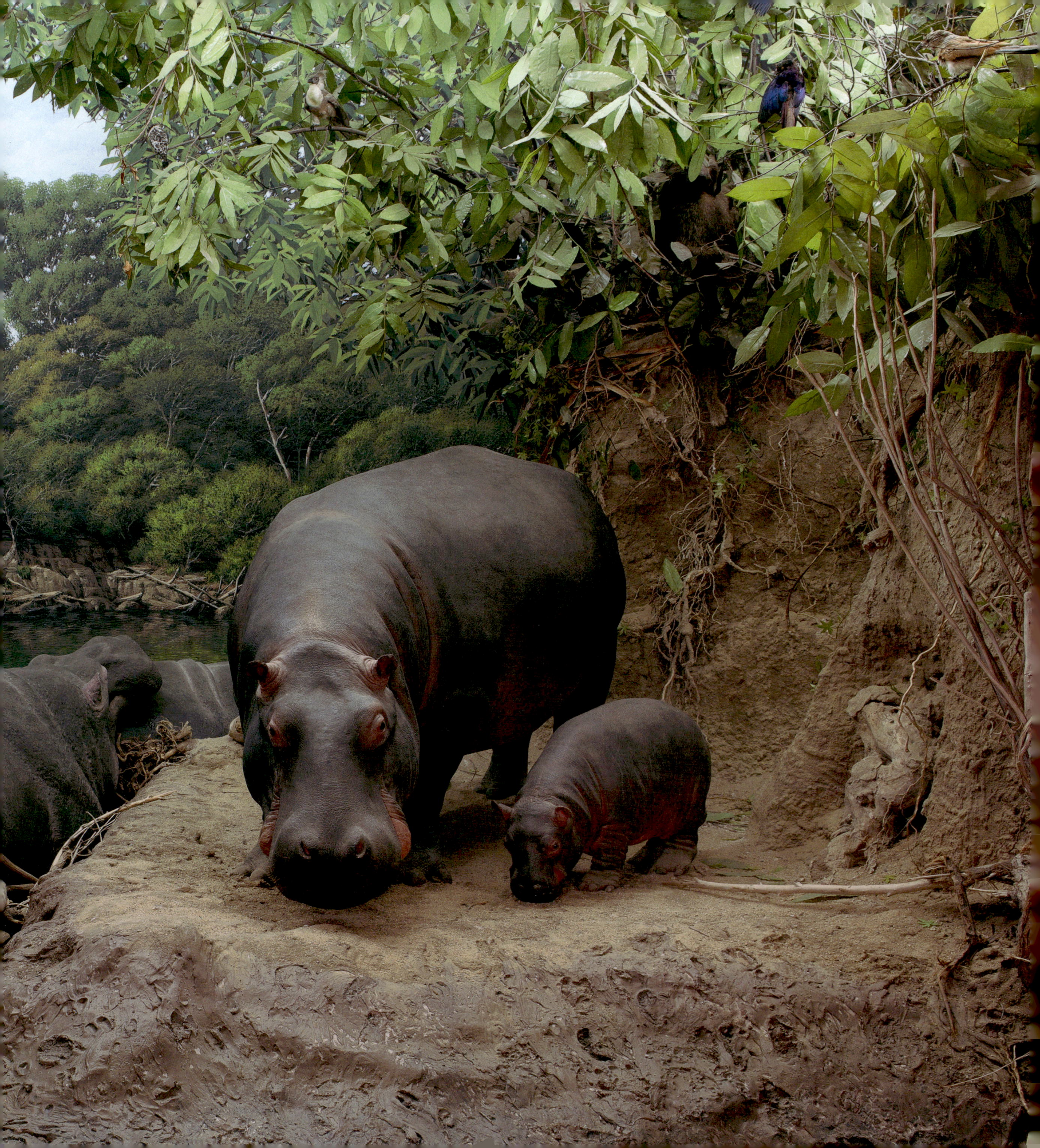

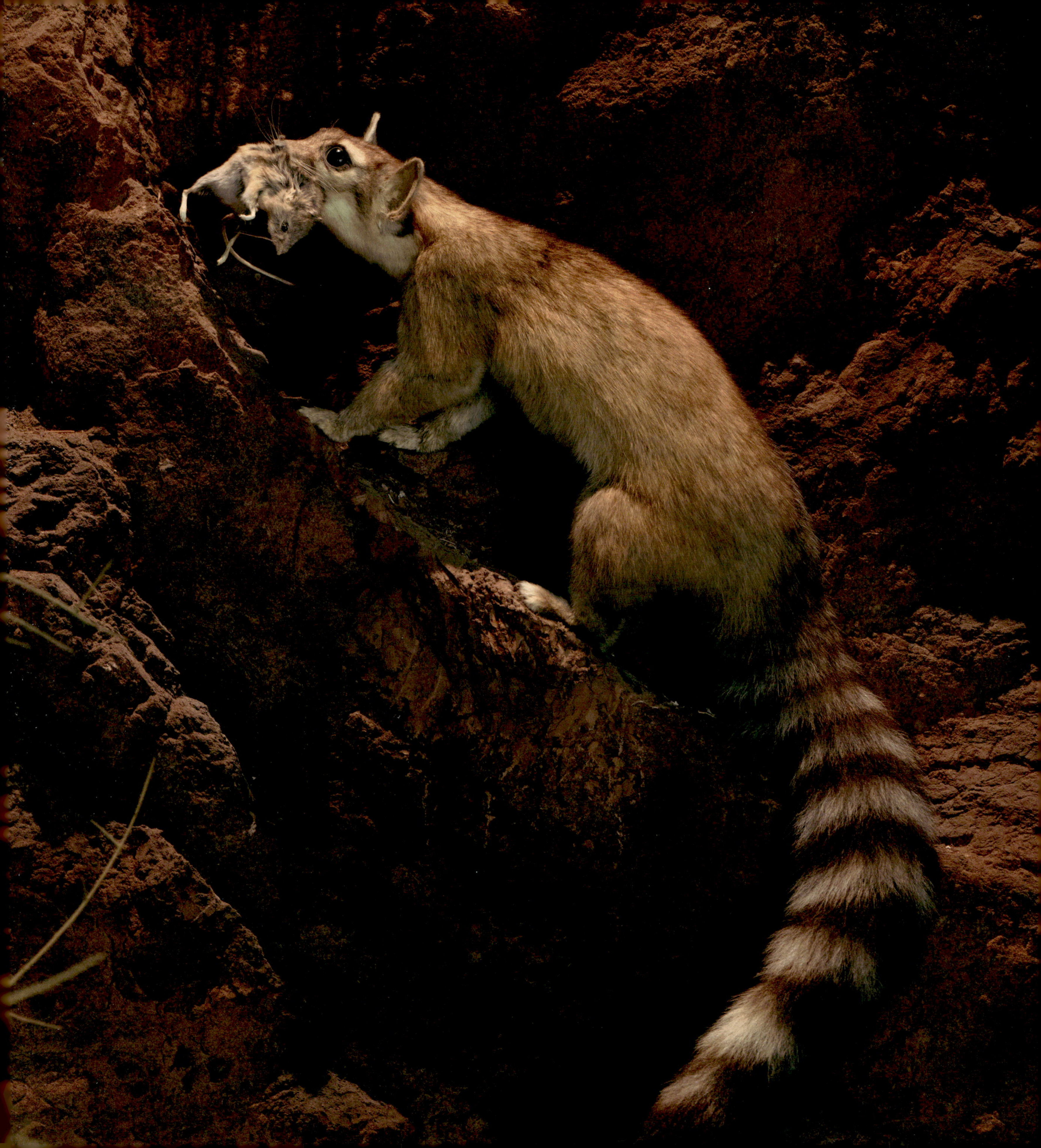

Composition

Matt Davis

Dioramas combine science and art. They are not exact duplicates of scenes found in nature, but rather accurate scenes designed first in the minds of artists. Long before any expedition, scientists, donors, and artists from NHM work together to plan out the diorama. Where should the location be? What is the season, time of day, and weather? Who are the focal animals and what are they doing? Artists start with quick sketches to work out the overall composition of the scene before building detailed miniature models called maquettes. These maquettes let artists experiment with different layouts, ensuring that all the parts of a diorama—from the background to the foreground to the taxidermy—work together to create a balanced, three-dimensional picture.

Although NHM's dioramas display habitats that range from deserts to rainforests, they tend to follow a standard format to educate and immerse viewers that was established a century ago. The scenes are set at ground level, so it feels like you are standing right there with the animals. The backgrounds, painted on curved canvas walls, create the illusion of distance and often illustrate famous mountain ranges or canyons found in national parks. The detailed foregrounds show the actual terrain the animals live in and the plants or other animals they eat. Early diorama artists openly admitted that they often

Previous spread
Dioramas like the *Hippopotamus* group are works of art created by a team of background artists, foreground artists, and taxidermists, collaborating to produce a composition that inspires and educates visitors about animals and their natural habitats.

← Museum Director Herbert Friedmann compared dioramas like this *Ringtail* group to great works of portraiture: in both, carefully planned compositions help viewers quickly realize the personality of the subject (2021).

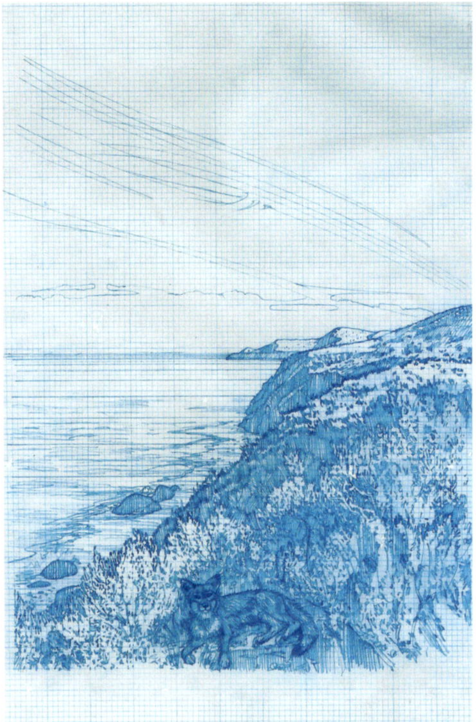

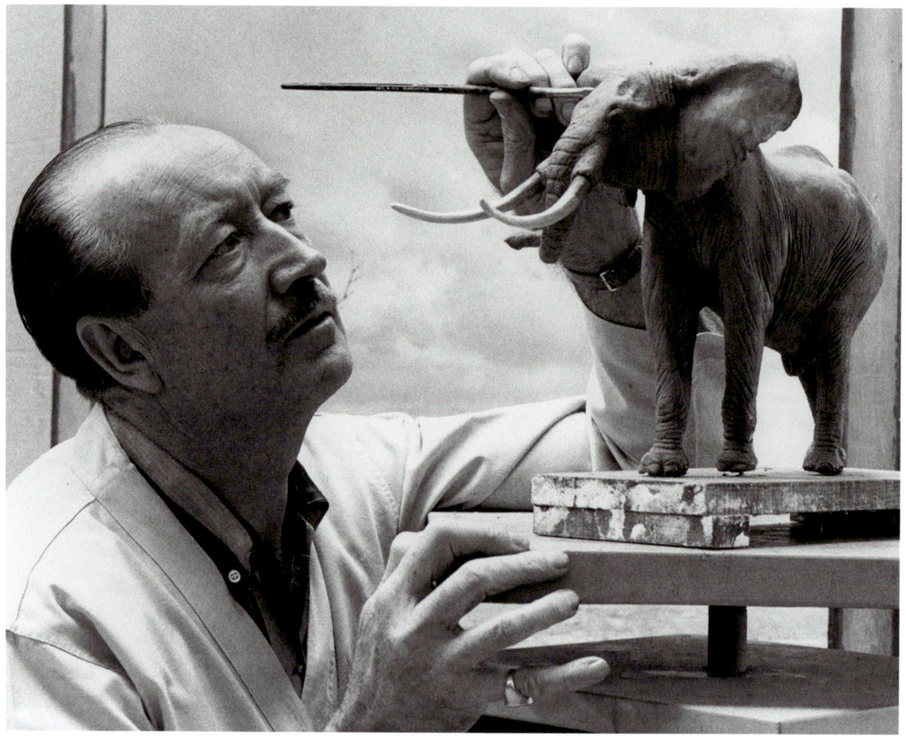

arranged taxidermied animals into unrealistic nuclear family groups similar to those of humans so that visitors could compare males, females, and juveniles in one place. "We like to have a full family represented in each of our groups," Senior Curator of Habitats Melville Lincoln once told reporters, "but we had to admit that bears really don't have a conventional family life."[23] (In many species of mammals, the male does not hang around after mating to help the female raise their offspring.)

Using their maquettes, artists could plan out groups so that "the general sizes, shapes, ages and structures of the animals were decided upon even before the expedition left Los Angeles."[24] These details were needed so that in the field, hunters would search only for animals that exactly matched the diorama's plan. As NHM's first taxidermist, Joseph P. Herring, wrote after the 1919 Alaska Research Expedition, "To kill game just to kill was not our motive. We killed only what we needed for groups."[25] Once the team came back from the field with reference photos, plant material, and animal hides, the real work of making a diorama began.

↖ Although many details changed from artist Robert Reid's initial sketch, the seaside setting and recumbent fox made it into the final *Island Fox* diorama (c. 2007).

↑ Taxidermists like George Adams often sculpt miniature models called maquettes to plan out the overall composition of new dioramas (1965).

→ Joseph L. Roop used his skills in stop-motion animation to build three detailed maquettes for the museum's first *Watering Hole* group, each exploring different background, foreground, and taxidermy layout options. These two maquettes show compositions that were rejected and never built (1929).

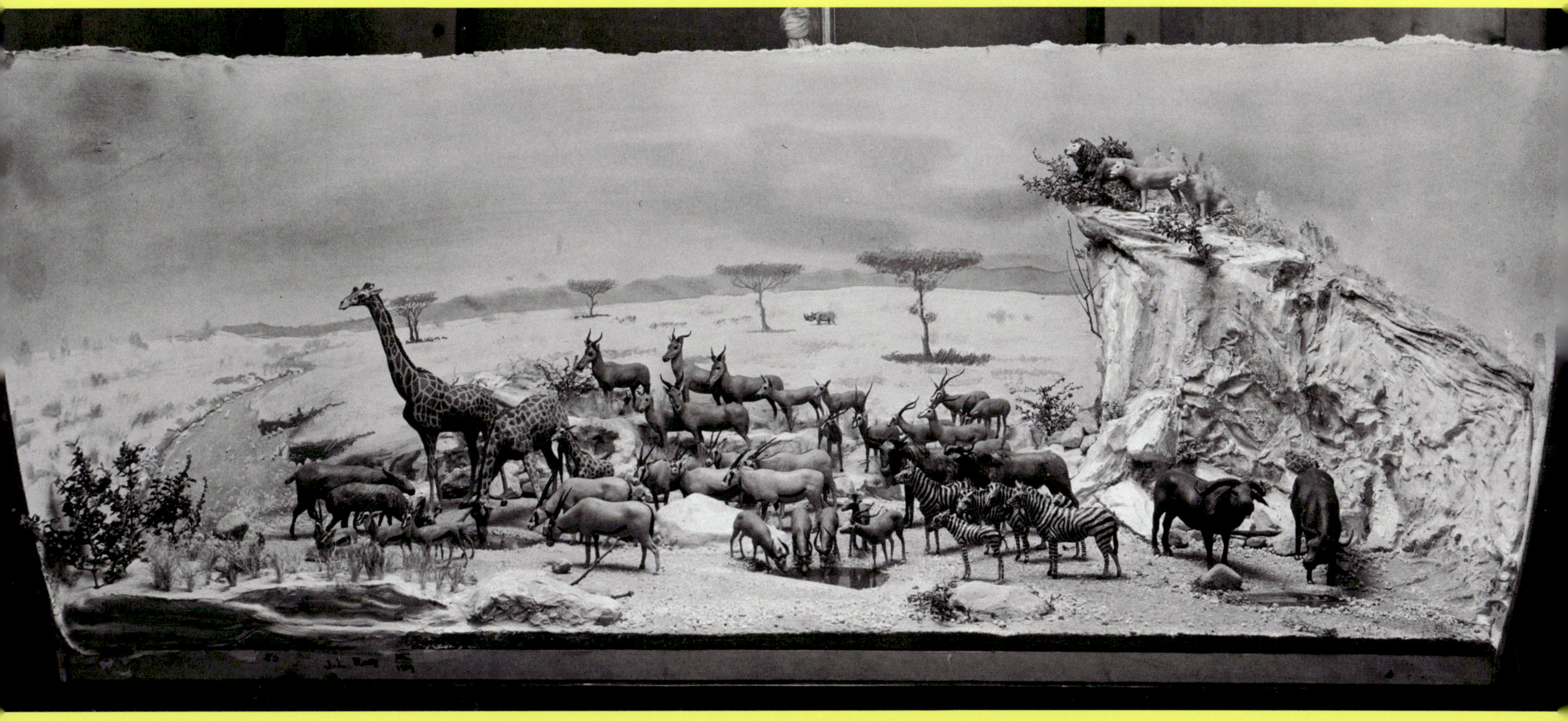

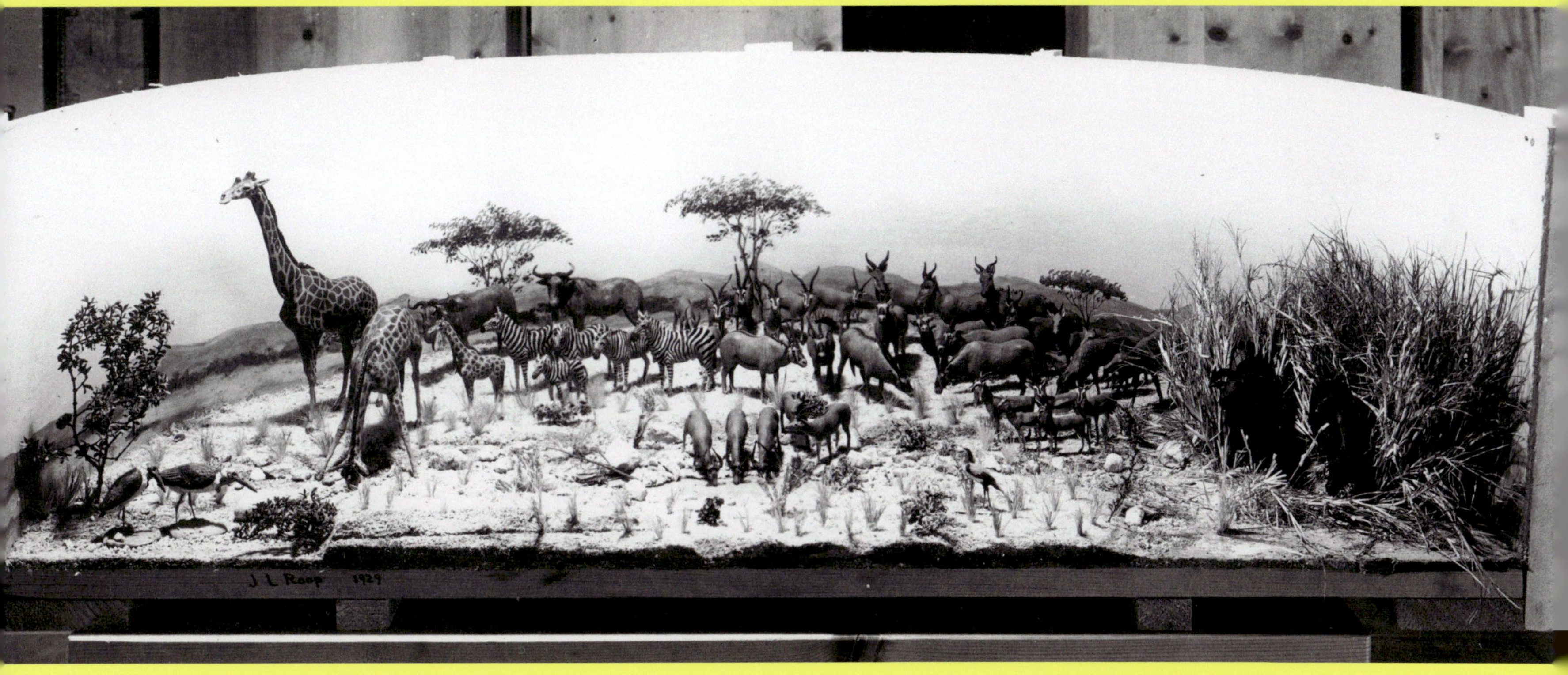
J L Roop 1929

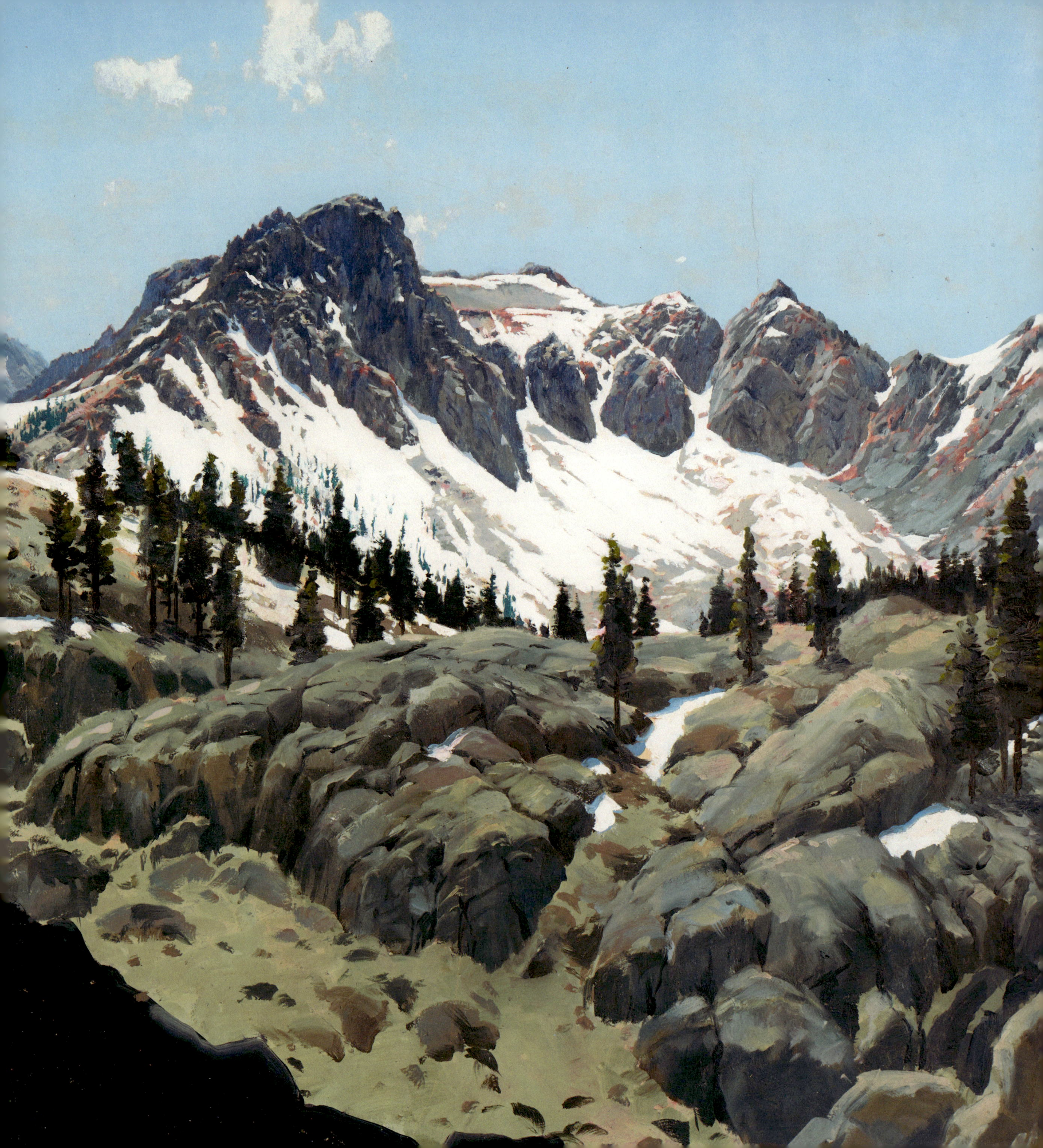

Backgrounds

Karen L. Maness

← Background artists use small-scale paintings, like this 1970s study for the *Fisher* diorama by Duncan Spencer, as a tool to test color, composition, and painting approaches before beginning the full-scale background murals (2023).

Backgrounds are integral to establishing the geography, season, and setting for each diorama. The background painter's command of the material, brushwork, and illusion bend curved walls into endless vistas. Every element of the background painting supports the diorama's narrative and educational purpose; each detail is considered for its visual and scientific balance and impact.

Background artists face a host of challenges in creating a convincing illusion. They must effectively render the represented location's atmosphere, time of day, colors, temperature, and quality of light, all on a gigantic canvas that can stretch over 75 feet (23 m) long. Furthermore, the artists must have an understanding of geological formations, botany, and atmospheric perspective, as well as the ability to render them convincingly. The illusion will break if the artist fails to tie the compositional elements, the horizon line, and the color value together with the 3D objects in the foreground.

Most NHM background artists did not have the luxury of traveling with museum expeditions to the locations depicted in the dioramas. They had to work from black-and-white photographs, color slides, film footage, and their personal knowledge to create a realistic setting. For some of the North American dioramas, though, artists have been fortunate enough to travel into the field, studying each location's atmospheric conditions, color palette, and principal elements, often gathering soil samples, plant specimens, and other items for reference.

After researching the desired habitat, either in person or through materials brought back by the expedition, the artist composes a drawing, arranging elements to create a poetic and scientifically grounded illusion of the location. The artist must consider the viewer's position; structural and architectural elements that may need to be hidden (such as support beams or columns); and the diorama bay's curved wall, designed to support the illusion of an endless vista. For example, backgrounds look most realistic when the horizon line is at the viewer's eye level. The *Watering Hole* diorama's horizon was set at 4½ feet (137 cm) off the ground, at eye level for the many thousands of children that visit the museum each year.

Using models, concept drawings, and paintings, the artist translates the location information into a functional design, with the foreground animals anchoring the structure and composition of each diorama. Once the composition is decided, the artist creates a more detailed painted sketch, devising the color palette and plans for lighting, referring to soil samples, plant specimens, and photographs gathered on location.

In the late 1930s, scenic artists employed in Hollywood filmmaking devised new painting systems to simulate reality on screen. A critical aspect of painting for motion pictures was speed, with scenic artists creating stunning large-scale environments in a fraction of the time required by traditional mural painters. NHM employed multiple Hollywood scenic artists, and many of their techniques were carried into the art of diorama painting.

Diorama artists work from the background to the foreground, beginning with the sky and then adding layered clouds, the distant background, the midground, and finally the foreground of the painting. These elements would be sketched on the wall in advance in a simple "cartoon." Scenic artist Duncan Alanson Spencer, who

↙ Staff Artist J. Robert "Bob" Sewell (1916-2008) paints a location study of the Fraser River Canyon in Canada for the *Gray Wolf* diorama in 1951.

↓ In the final *Gray Wolf* background, Sewell shifted the season to winter and covered all evidence of the canyon's riverside road underneath a painted herd of elk, revealing that the goal of the background artist was "not a photographic rendition, but an illusion of reality" (2014).[26]

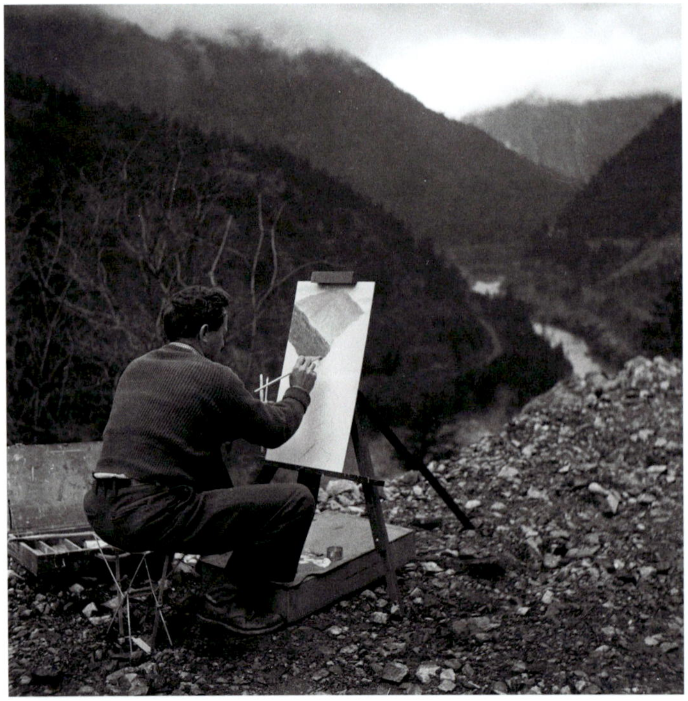

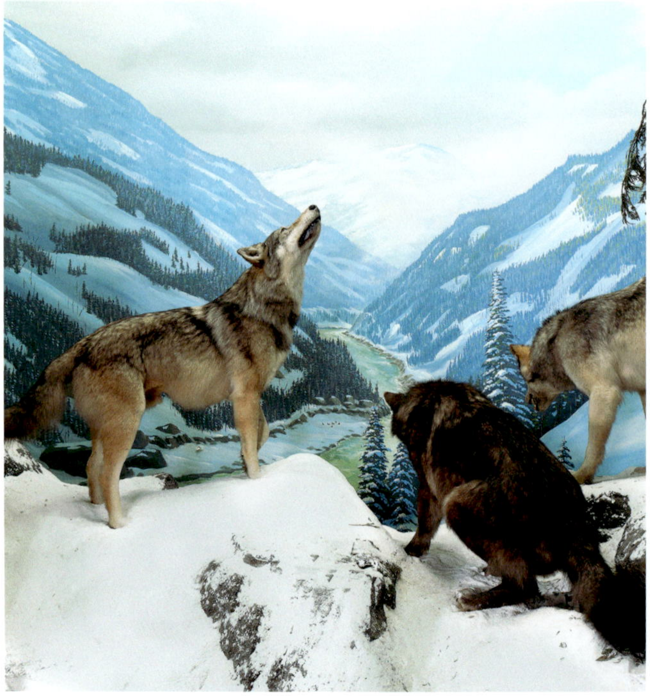

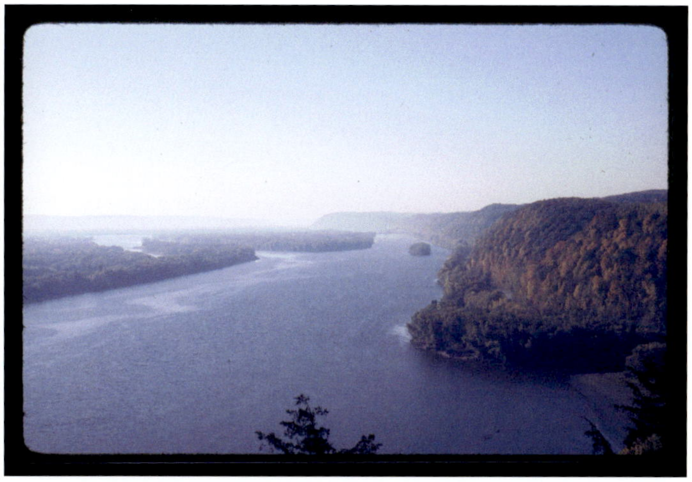

← A color reference slide of the Mississippi River near McGregor, Iowa, used by Duncan Spencer to paint the *Red Fox* diorama background (1975).

↓ Spencer's final background for the *Red Fox* group, painted around 1977, features different atmospheric conditions and a setting a couple of weeks earlier in the fall. It also created a more dynamic composition by displaying the water channels from a higher vantage point (2021).

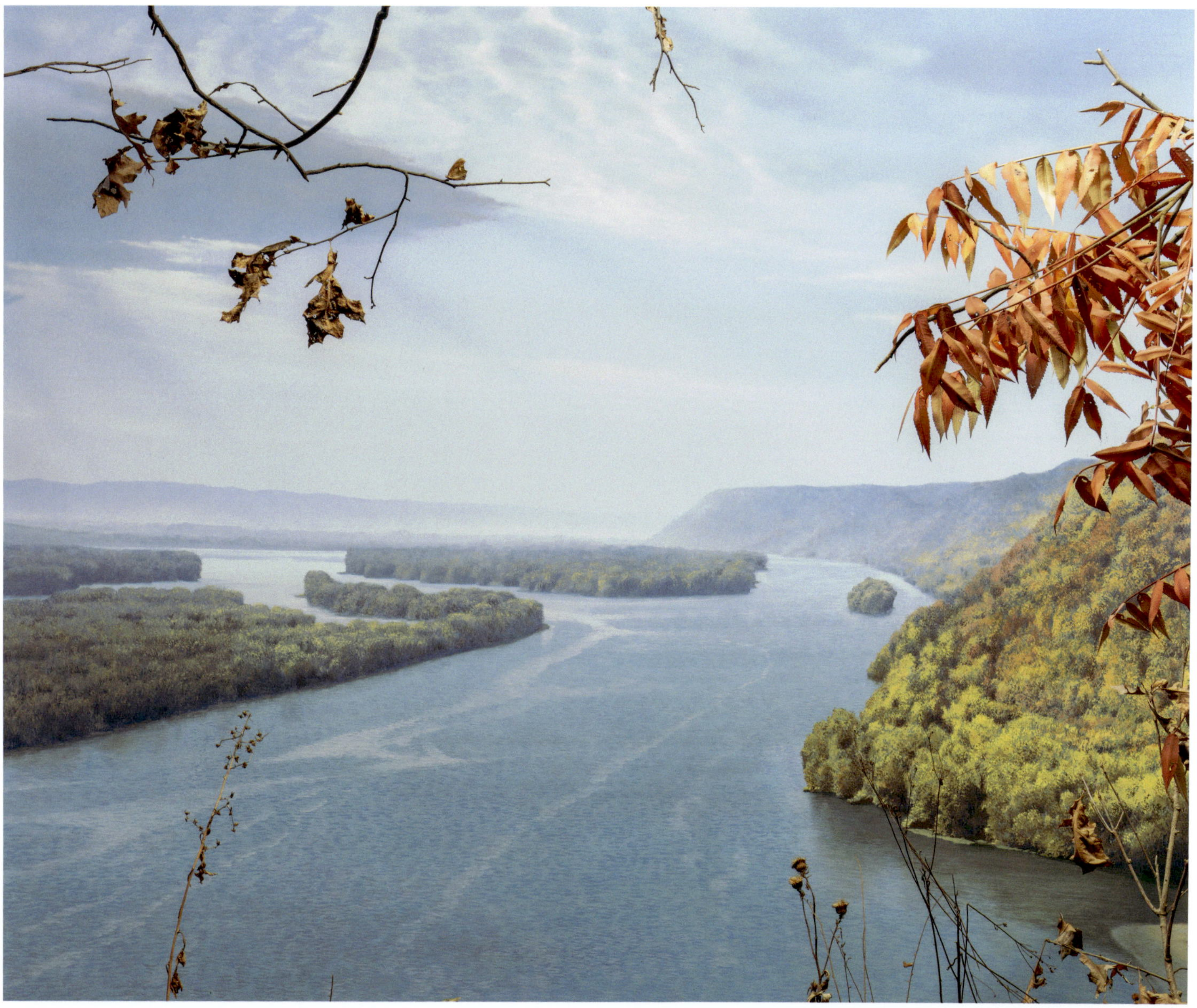

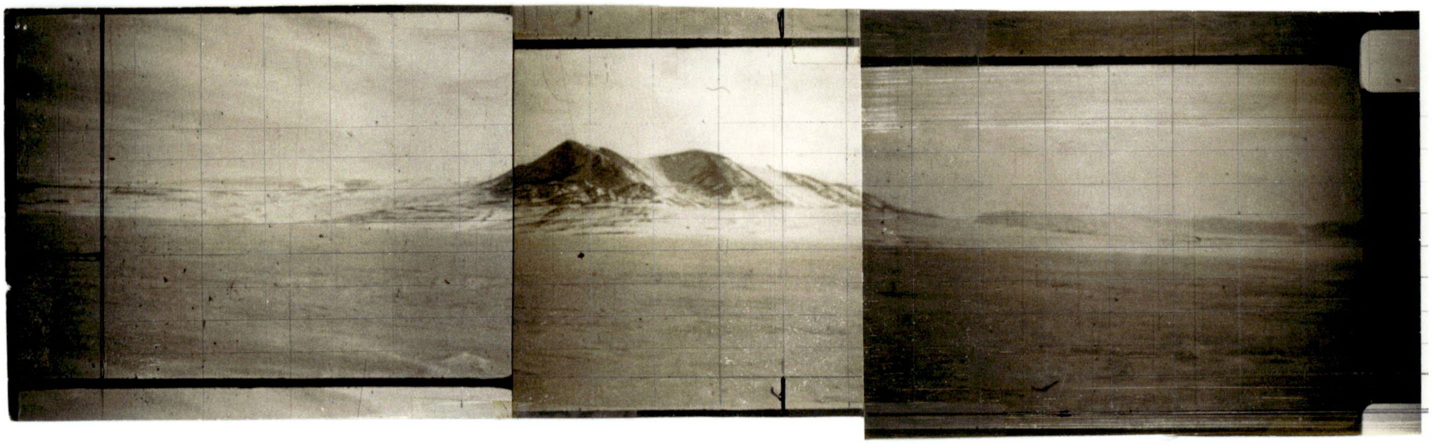

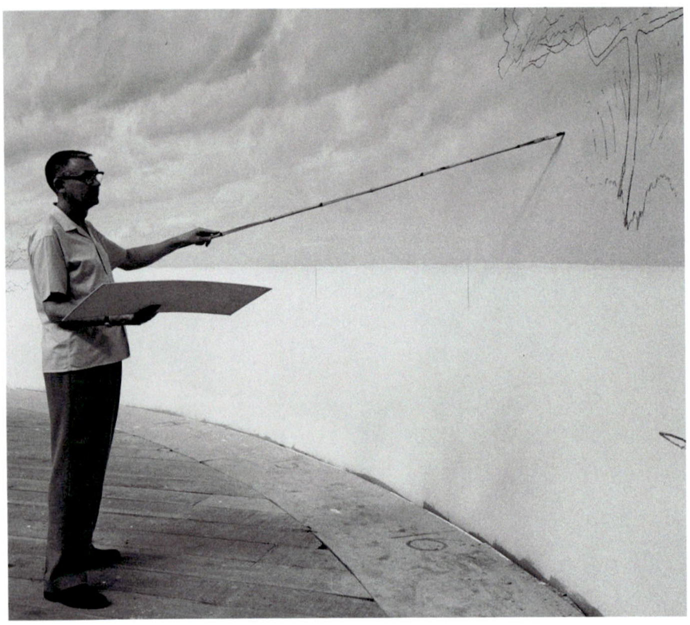

← Spencer uses a bamboo extension pole to "cartoon" the background tree line and middle distance details of the *Watering Hole* diorama. He holds a gridded sketch that corresponds with the faint vertical lines coming down from the horizon line of the diorama (1965).

↑ Since he wasn't part of the original 1960 Logan Arctic Expedition, Spencer took a film shot on Ellesmere Island, enlarged the frames, and taped them together to create a panoramic reference for version 2 of the *Muskox* group. The grid allowed him to accurately transfer this scene to the large background of the diorama bay (c. 1976).

had worked on over 400 Hollywood movies, used a traditional scenic art tool called an extension, a thin bamboo rod tipped with vine charcoal, to draw the cartoon. Such extensions allow artists to view their work at a distance as they draw, using long strokes. Control and command of drawing were obviously absolute musts. Consulting a gridded scale drawing, artists would make vertical marks on the wall, creating a loose grid to ensure an accurate translation of the smaller scale drawing to the much larger diorama.

Hollywood scenic artists made custom brushes and tools from tufts of hair, carpet rollers, bundles of cloth, and stamps to speed up the painting process for motion picture backgrounds. Textured rollers simulated bark, and bunches of pipe cleaners bent like blades of grass created endless grasslands. Artists also modified pneumatic spray guns to allow them to paint optically blended spattered-color sky gradations and soft clouds. At NHM, you can see the result of such techniques in Spencer's *Serval*, *Stone's Thinhorn Sheep*, and *Polar Bear* dioramas, which exemplify expert cloud representation, color use, and composition, successfully marrying the background and foreground elements.

In 1964, Director Herbert Friedmann explained the goal of background painting:

> In order to make the [diorama] as communicative, as informing as possible, no sign of the human effort involved in its construction should intrude. The artist painting the background has to forget "personal style" and to create, not a photographic rendition, but an illusion of reality.[27]

Although the background paintings remain unsigned and undated (at least where visitors can see), each reveals its creator's hand through their aesthetic preferences, painting approaches, and stylistic influences. Sophisticated brushwork of leading oil painters dominated the earliest dioramas in the museum. And in each generation of the museum's 100-year evolution, examples of exceptional work emerge. The halls' most prolific and celebrated artists include Hanson Puthuff, Frank J. MacKenzie, and Duncan Spencer.

Award-winning Missouri-born Hanson Duvall Puthuff (1875–1972) lived a rough frontier life before studying at the University of Denver Art School. For years, he worked as a commercial painter and billboard artist, moving out to Los Angeles in 1903 to paint for an advertising company. He quickly grew enamored of the area, becoming well known as a California Impressionist of the Eucalyptus School for his plein air landscape paintings. Characterized by loose brushwork and bright colors, his work captures the effects of light and atmosphere. As a well-regarded local pleinairist, Puthuff was frequently invited to exhibit at the museum, which was

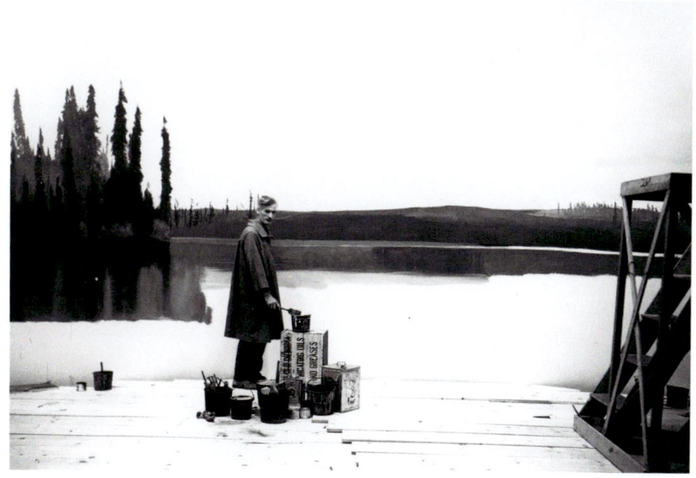

↑ Hanson Puthuff applies the initial color values to the *Moose* group background, working loosely in this initial phase. Note the confident brushwork on the treetops and the blended reflections of those treetops in the lake (1926).

↓ A detail of the *Mule Deer* background painted by Puthuff around 1927. Ultramarine shadows create depth and contrast between the background and foreground (2021).

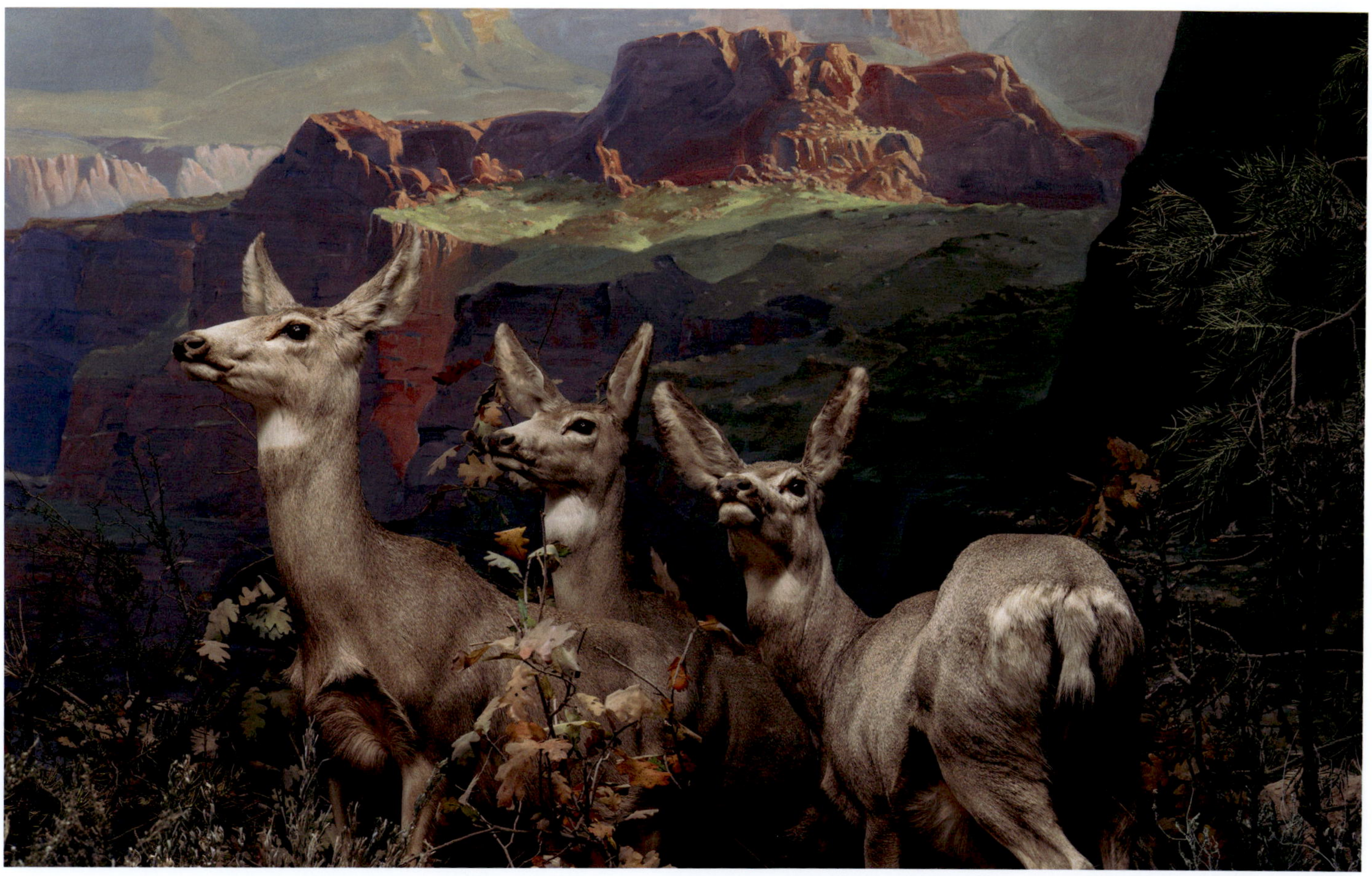

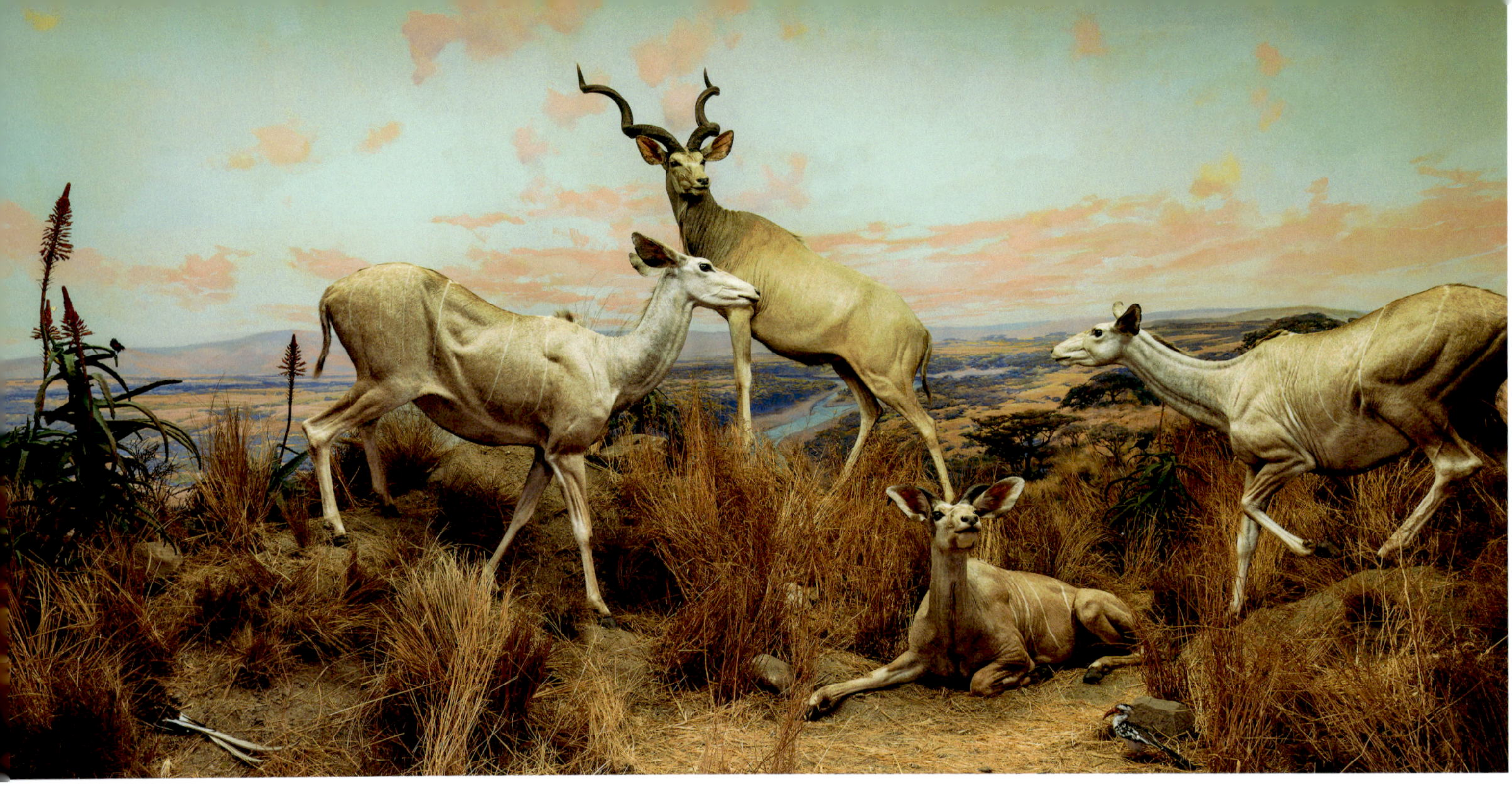

↑ The majestic *Greater Kudu* painting by Frank MacKenzie frames the central figures in a sweeping, romantic background. Although he painted all of the original backgrounds in the Simson Hall of African Mammals (now the Oschin Hall of African Mammals), only two of his backgrounds survive (2014).

→ Staff Artist Robert Clark (1920–1999) consults a small background study while painting the first version of the *Hippopotamus* diorama around 1959. Clark later painted in grass below his knee to tie the 2D background into the yet-to-be-constructed 3D foreground.

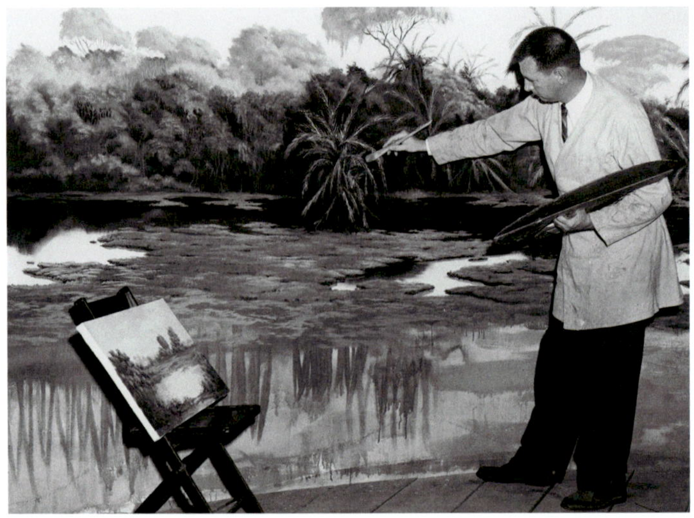

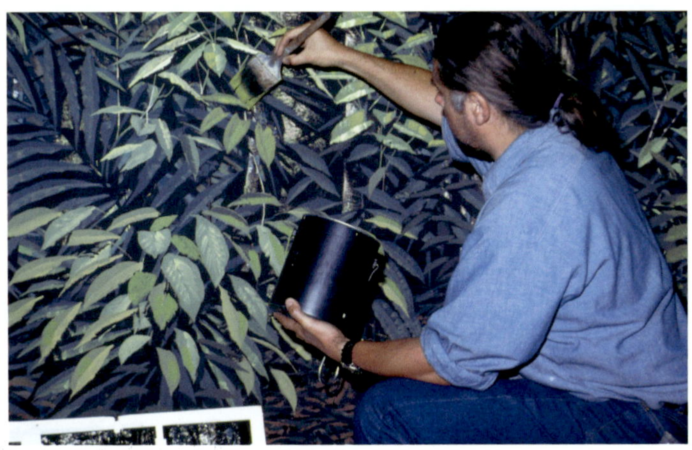

→ Staff Artist Robert Reid uses bright, warm-colored leaves painted on top of darker, cooler leaves to create an optical illusion of depth in the *Western Gorilla* group (c. 1999).

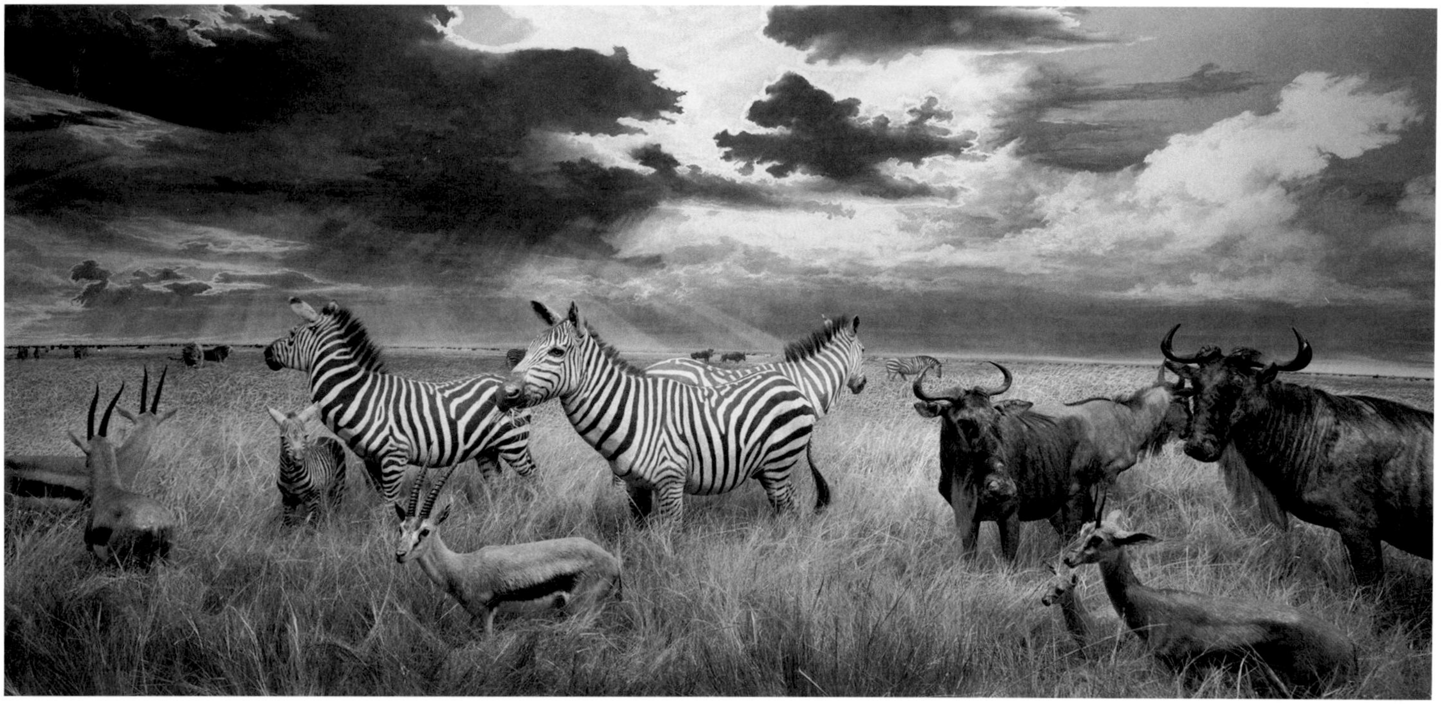

↑ An interesting way to check an artist's ability to create depth is to filter out the color. The image should read as convincingly in black and white as in color, as in this photo of Larry Latta's *Serengeti* group background (1965).

then known as the Los Angeles County Museum of History, Science *and Art*. (In 1961, much of the art moved to LACMA, the newly created Los Angeles County Museum of Art.) During the 1920s, Puthuff painted most of the original diorama backgrounds in the Level 1 Hall of North American Mammals, such as the *Mule Deer*, *Caribou*, and *Mountain Goat* groups. Puthuff's diorama backgrounds mirror his great works of art hanging in museums and private collections and merit recognition as monumental original artworks of their era.

In 1929 the award-winning London-born artist Frank J. MacKenzie (1865–1939) was hired to paint all the backgrounds for the original Simson Hall of African Mammals, although only two of his backgrounds survive in what is now the Mr. and Mrs. Samuel Oschin and Family Hall of African Mammals. The hall's original patron, Leslie Simson, had already commissioned artist Maurice Logan (1886–1977) to travel around Africa with him and produce site studies for the animals Simson was collecting. Having served in the military in colonial Africa for many years, MacKenzie knew African landscapes well and was able to adapt Logan's studies into full-scale backgrounds at incredible speed, finishing the hall in only seven months. He likely accomplished this with the capable assistance of his artist wife, Florence. Florence MacKenzie (1890–1968) studied art at the Corcoran Gallery of Art, the National School of Fine and Applied Arts, and the Philadelphia School of Design for Women. She is credited

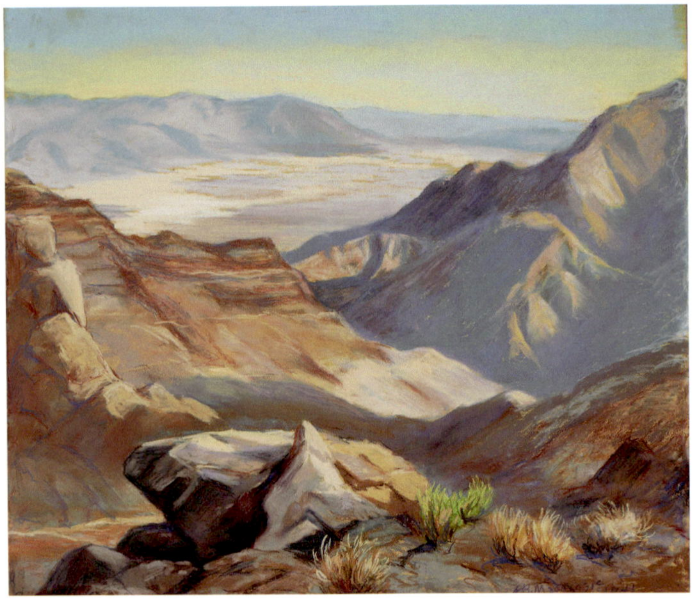

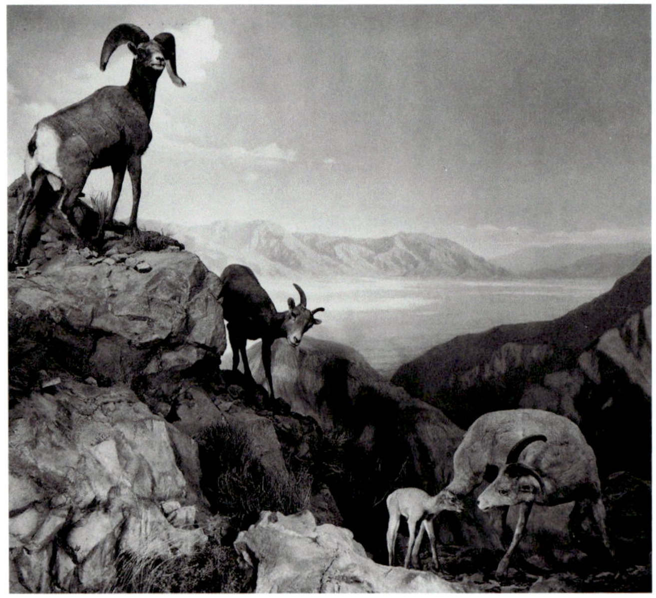

↖ Florence MacKenzie's 1941 *Bighorn Sheep* background concept sketch (2023).

↑ Florence MacKenzie's completed *Bighorn Sheep* diorama, about 1944, varies from her original conceptual sketch. She reversed the shadows, mountain structures, and light direction, probably to accommodate the diorama's lighting positions and draw focus to the figures in the foreground (c. 1970s).

with two backgrounds at NHM, both of which are examples of later artists painting over diorama backgrounds, a common occurrence at the museum.

In 1941, two years after Frank died, Florence MacKenzie was hired to paint the *Bighorn Sheep* habitat. She accepted the low sum of $350 for the work due to her precarious financial situation as a widow and social norms that devalued women's talents and labor. Duncan Spencer (1911–1988) was hired to repaint the group's background in the 1970s to make it more realistic. His foreground composition, background, and color choices resemble Florence MacKenzie's original pastel sketch, a nod to her expertise as an artist. MacKenzie herself painted over others' backgrounds. In 1944, she covered parts of her late husband's *Reedbuck* background to create the new *Spotted Hyena* group.

Spencer's first diorama painting was the second version of the museum's *Watering Hole* group, which featured a whole family of elephants. He began in 1965, using imagery from the prize-winning color documentary *Safari Ya Tembo*, produced by museum patron and expedition leader Maurice Machris. Spencer had arrived at the museum in a period of significant personnel change and prosperity. A Southern California watercolorist and award-winning pleinairist, he felt confident with the subject, as he had often painted African landscapes in backdrops for Hollywood motion pictures.

Spencer had studied at the Chouinard Art Institute before beginning his 38-year-long career as a Hollywood scenic artist in 1937, painting over 400 film backdrops for Metro-Goldwyn-Mayer (MGM), Columbia, and 20th

→ In version 2 of the *Bighorn Sheep* group, Spencer's new background painting nods to Florence MacKenzie's earlier version by adopting her color palette and general composition. This painting features wider desert vistas, drawing forward the cool tones of the background (2021).

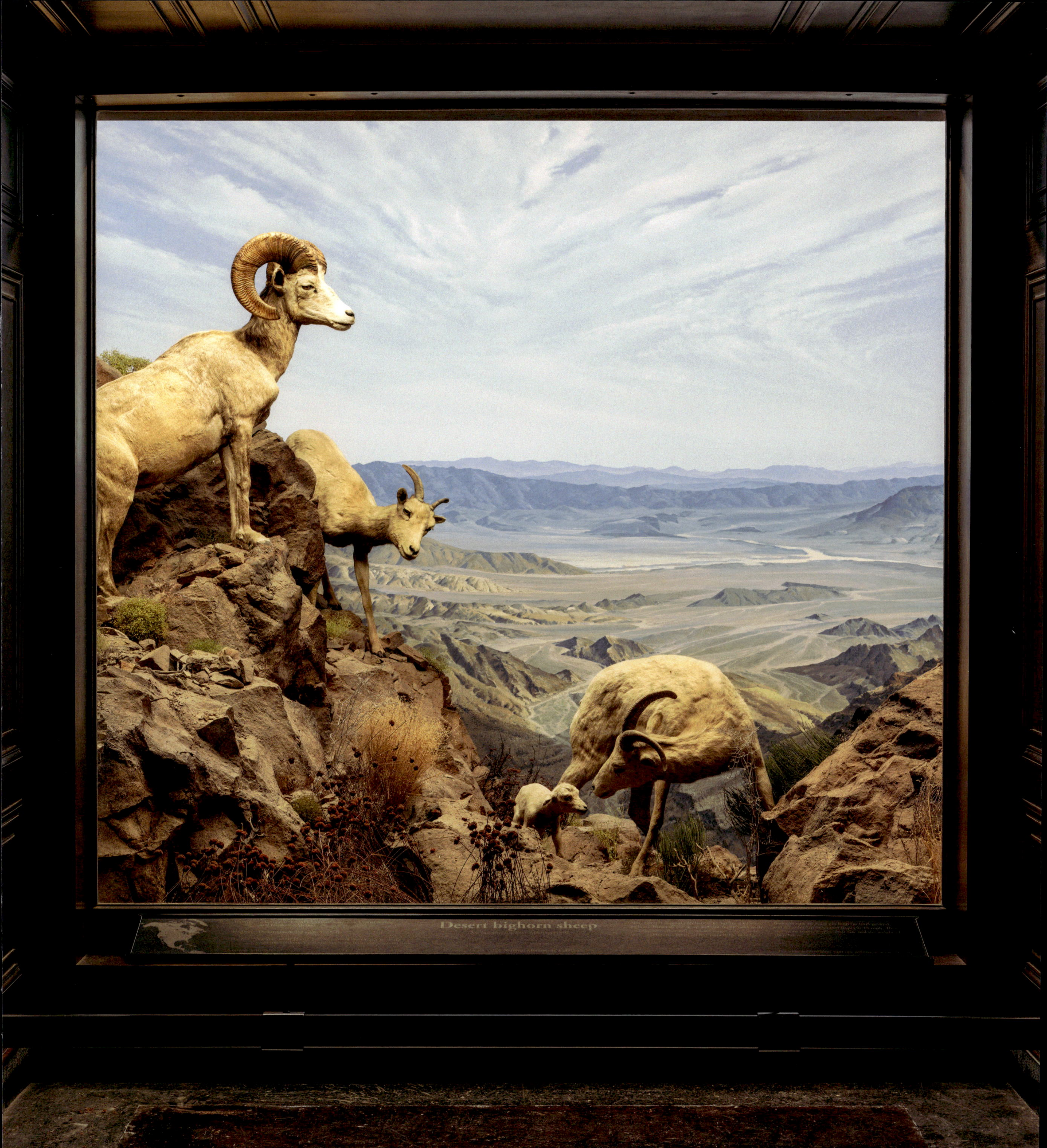
Desert bighorn sheep

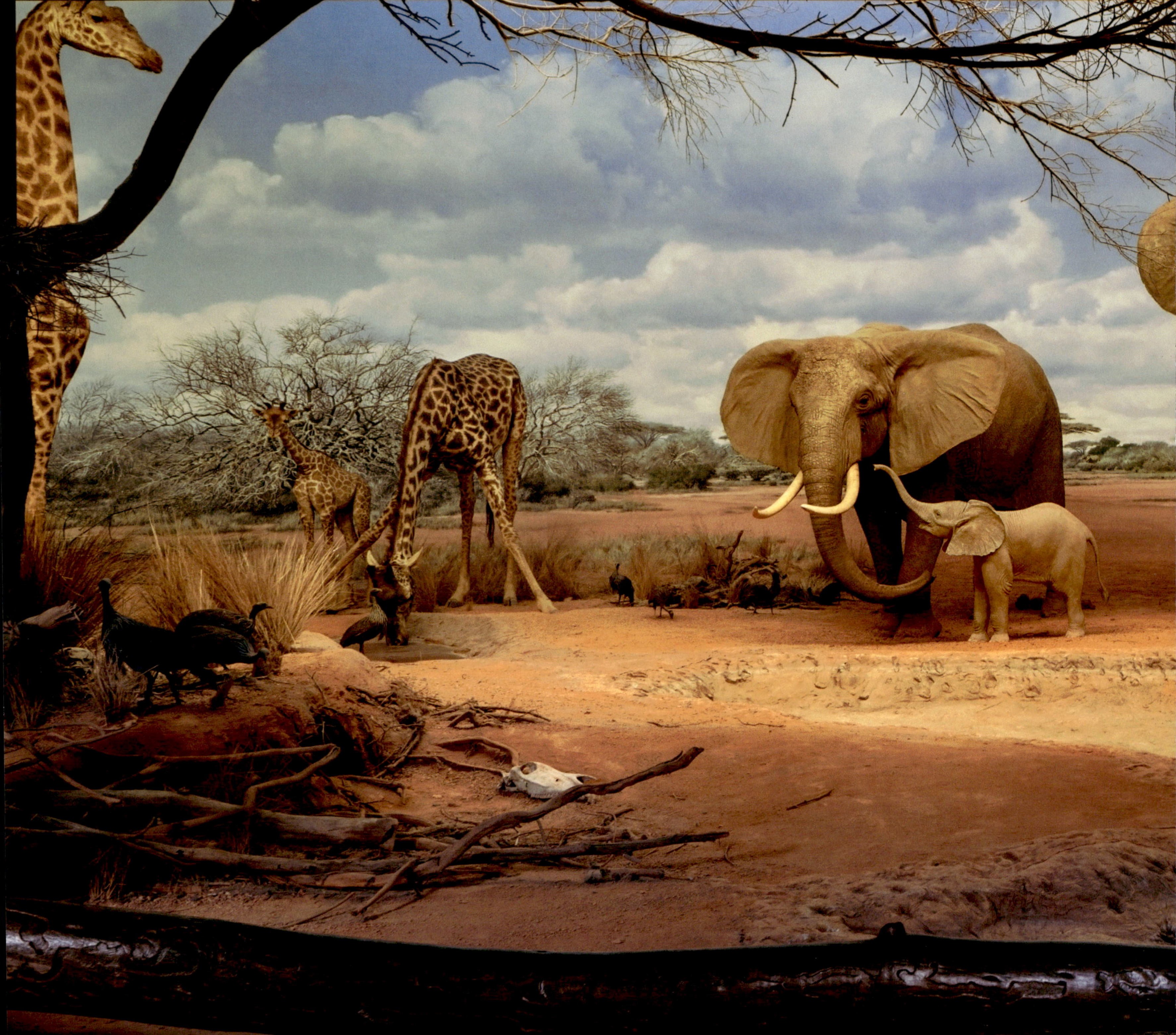

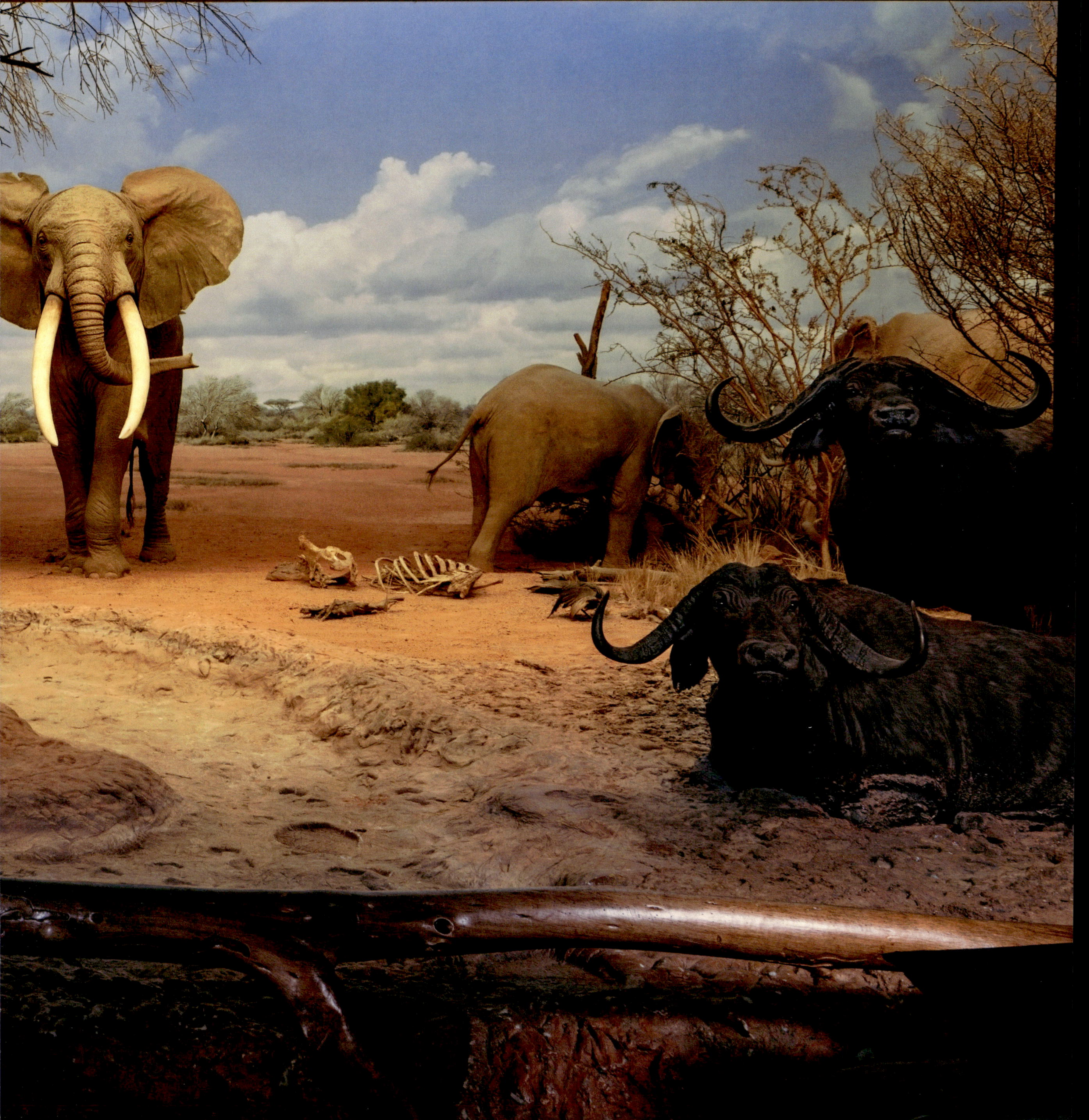

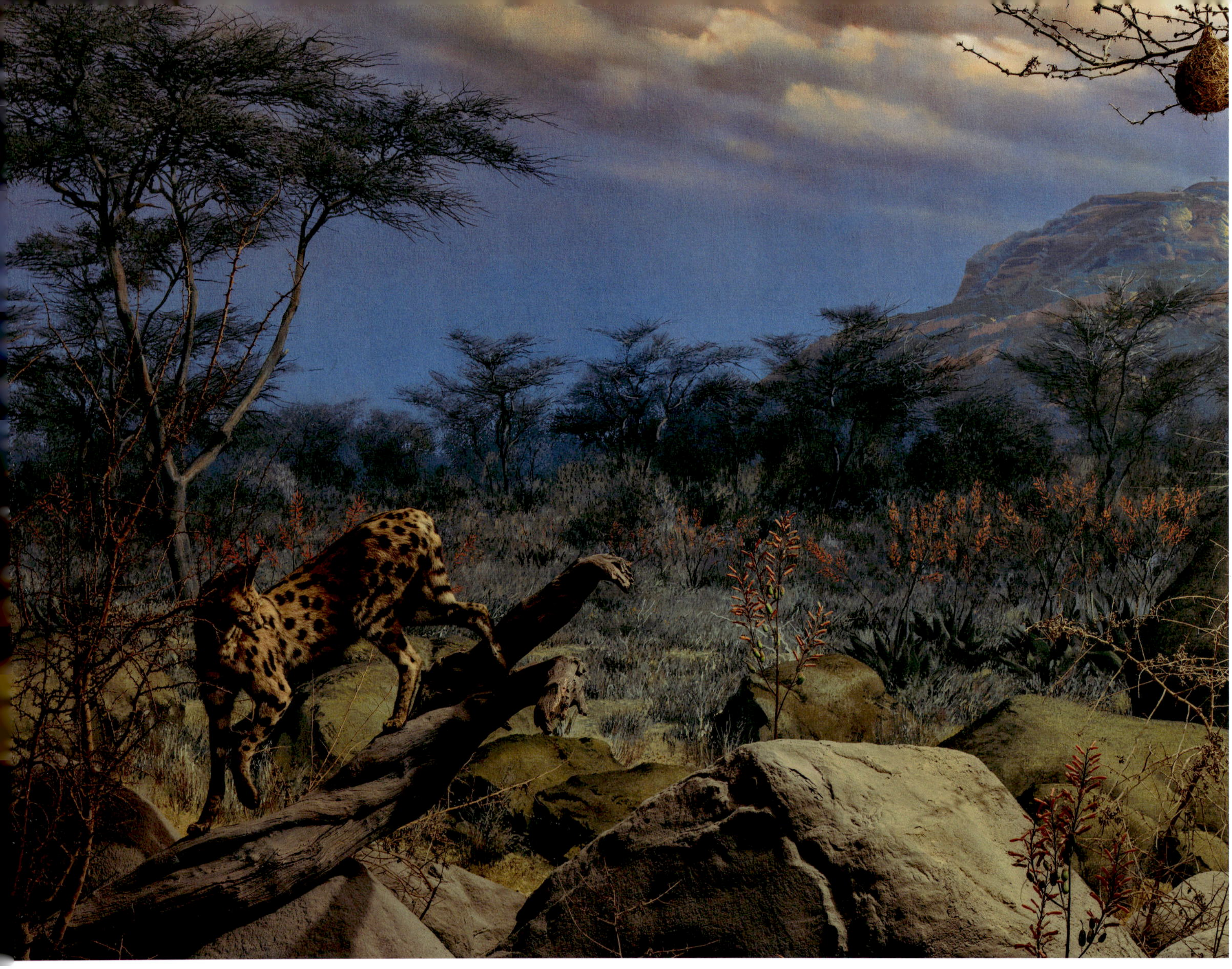

Century Fox. His film credits include *The Wizard of Oz*, *White Christmas*, *Singin' in the Rain*, *North by Northwest*, and *Mr. Smith Goes to Washington*. Diagnosed as color-blind, the artist credited his years in the motion picture industry with helping him overcome his color limitations. When the motion picture industry began, all backdrops were painted black and white, with varying shades of gray, because the films themselves were black and white. Spencer transferred his command of convincing value structures to full-color backdrops, becoming one of Hollywood's top scenic artists. He would eventually paint 19 habitats in the museum, working part-time year-round, assisted by Hollywood motion picture scenic artists Bob Oberbeck (1909–1988) and Gordon Butcher (1908–2001). Butcher recalled the *Polar Bear* diorama as one of their best. Many at the museum respected Spencer's

Previous spread
Duncan Spencer painted the second version of the *Watering Hole* group background, over 116 feet wide and 37 feet tall (approx. 35 × 11 m), in 1965, finishing in just six months. The background transports viewers to the open, arid landscape of southern Kenya, where African elephants, giraffes, and Cape buffalo roam. The artist carefully calculated the perspective and height of trees in the background so that the horizon appears 3½ miles (5.6 km) from a viewer standing in front of the diorama (2014).

↑ Spencer's dynamic and sophisticated *Serval* habitat is a quiet gem. Museum visitors who approach the intimate diorama and peer up are rewarded by a glowing skyscape optically vibrating off the cool-cast sky bathing the horizon (2021).

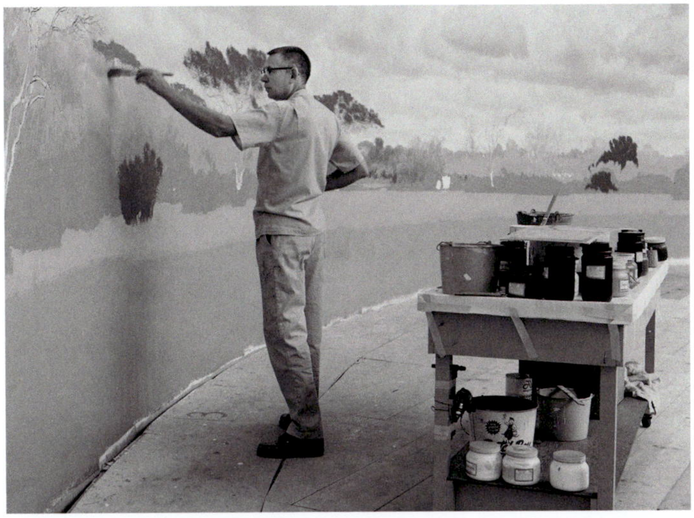

← With the sky of the *Watering Hole* group complete and broad foreground element shapes and values laid in, Spencer begins carving out the shadow shapes to define the forms of trees and foliage in the distance (1965).

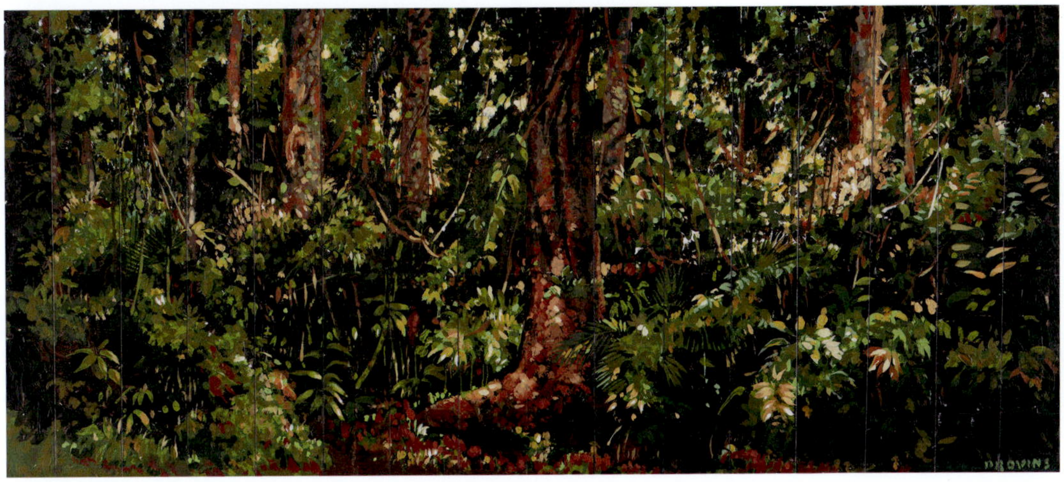

← Clark Provins likely painted this study for the *Okapi* group's jungle background in the early 1980s. Note the faint grid lines used for transferring the image accurately to the large curved wall of the diorama (2023).

mastery. Curator of Birds and Mammals Kenneth Stager wrote, "The habitat backgrounds of Duncan Spencer, I am thoroughly convinced, are without equal anywhere in the world."[28]

In 1967 Idaho-born oil painter, watercolorist, and lead MGM scenic artist Clark M. Provins (1909–1991) joined Spencer in designing and painting backgrounds, including the *Virginia Opossum*, the *Pronghorn*, and the *Okapi*. After the decline of the Hollywood studio system, Provins was one of the last artists in 1970s Hollywood capable of executing full-scale architectural scenic backdrops for motion pictures; he and fellow MGM scenic artist Oberbeck worked together to lead backdrop projects for Warner Bros. Provins worked late in life mentoring painters for film and television, influencing new generations of younger scenic artists.

Painted diorama backgrounds are quiet masterpieces that project the wonder and majesty of the natural world. Though often unnoticed, they play an integral role in the science and storytelling of habitat groups. Meant to function not as paintings but as "illusions of reality," the diorama backgrounds at NHM are rarely recognized as the great works of large-scale art that they are. Even when their art went unrecognized as such, the background painters were able to find greater meaning in their work. In a 1926 article about the dioramas he was then painting, Puthuff wrote, "I consider it a particular privilege to be identified with this very important undertaking, which will be valued and appreciated more and more as the years and generations pass, as the animals become extinct and their habitations are invaded by mankind."[29] It is a testament to the skill of NHM's background painters that the latter half of Puthuff's prediction about extinctions never came true. The dioramas these masters painted have evoked a reverence for wilderness that has helped save the animals inside them from disappearing forever.

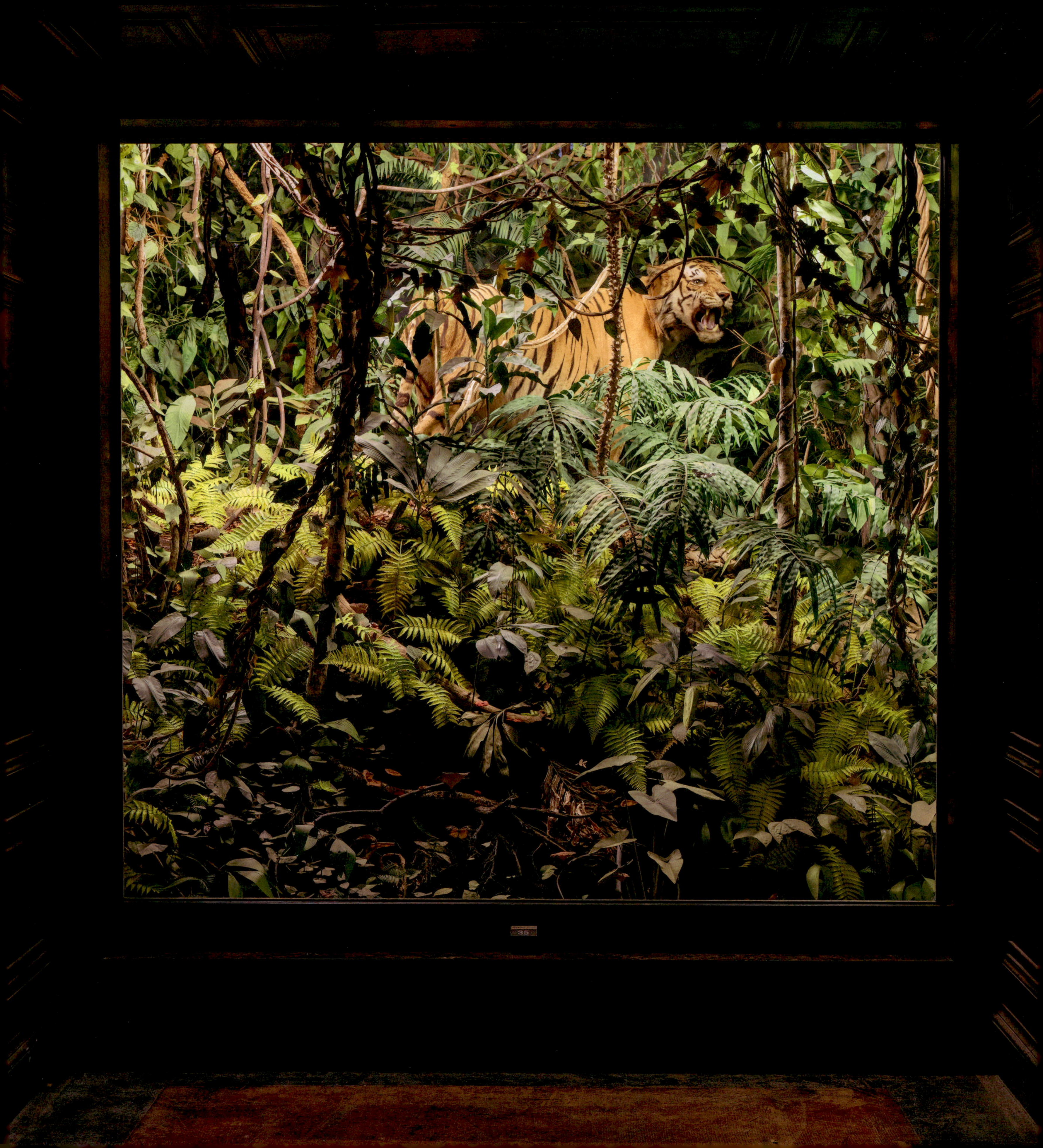
35

Foregrounds

Tim Bovard

← The *Tiger* diorama, opened in 1977, is almost completely foreground, with thousands of handmade leaves obscuring the taxidermy and background. Starting in the 1960s, new techniques like vacuum forming, as well as the addition of multiple skilled model makers to the staff, allowed the museum to produce dioramas with increasingly detailed and leafy foregrounds (2021).

The foreground of a diorama is the realistic three-dimensional landscape in front of the two-dimensional background painting. It represents the ground and may be composed of soil, rocks, plants (both real and artificial), and sometimes even water features. For the animals in the diorama, the foreground is their stage.

Foregrounds start in the field; artists use reference photographs, field notes, and molds made on location to produce a realistic scene later at the museum. The foreground, in concert with the background, should create the illusion of a specific season and a specific time of day in a real geographic location.

Ideally, the background and foreground design decisions for a diorama will be made at the same time, so that the visual tie-in between the 2D background and 3D foreground can be planned seamlessly. This is especially important for backgrounds and foregrounds that connect closely with no major space or drop-off between them. The tie-in elements may be rocks, grass, leaves, or other objects. For the *Gray Fox* group, the sandy ground the foxes fight on extends directly into the San Gabriel Mountains behind them, tying the foreground and background together. Any elements like plants or animals that go directly from foreground to background need to be produced early on so that the background painter can make sure their painting matches the coloration and appearance of the actual 3D elements. In dioramas like the *Pacific Marten* group, the visual tie-in between foreground and background is a steep drop off a cliff, which creates the illusion of space and distance. Decisions concerning lighting are also essential to the design of the tie-in. Generally, a slight gap must exist between the plants in the foreground and the background painting in order to avoid shadows falling onto the background. Lighting is also adjusted to help create the illusion of a certain time of day or season.

The plants in dioramas can be real dried ones or replicas made from molds of real leaves. Both kinds have their start out in the field. Foreground artists travel to

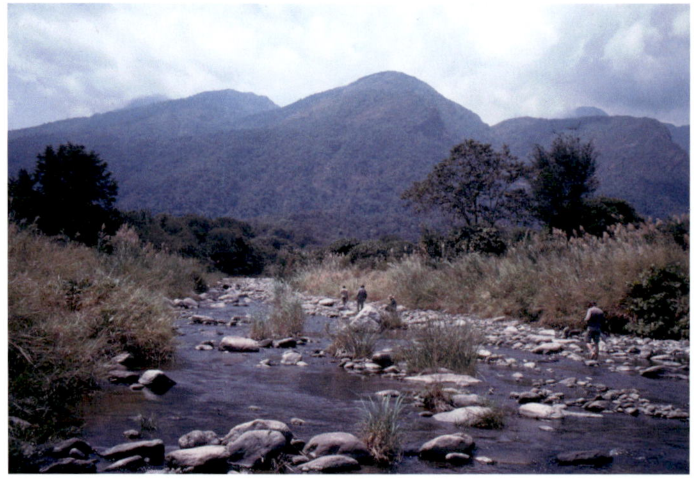

↗ Foreground artists traveled to the Mahale Mountains of Tanzania on the 1989 Thornton Snider Expedition to study the landscape and collect plant molds for version 2 of the *Chimpanzee* group, which reused existing taxidermied chimps from an earlier 1970 diorama.

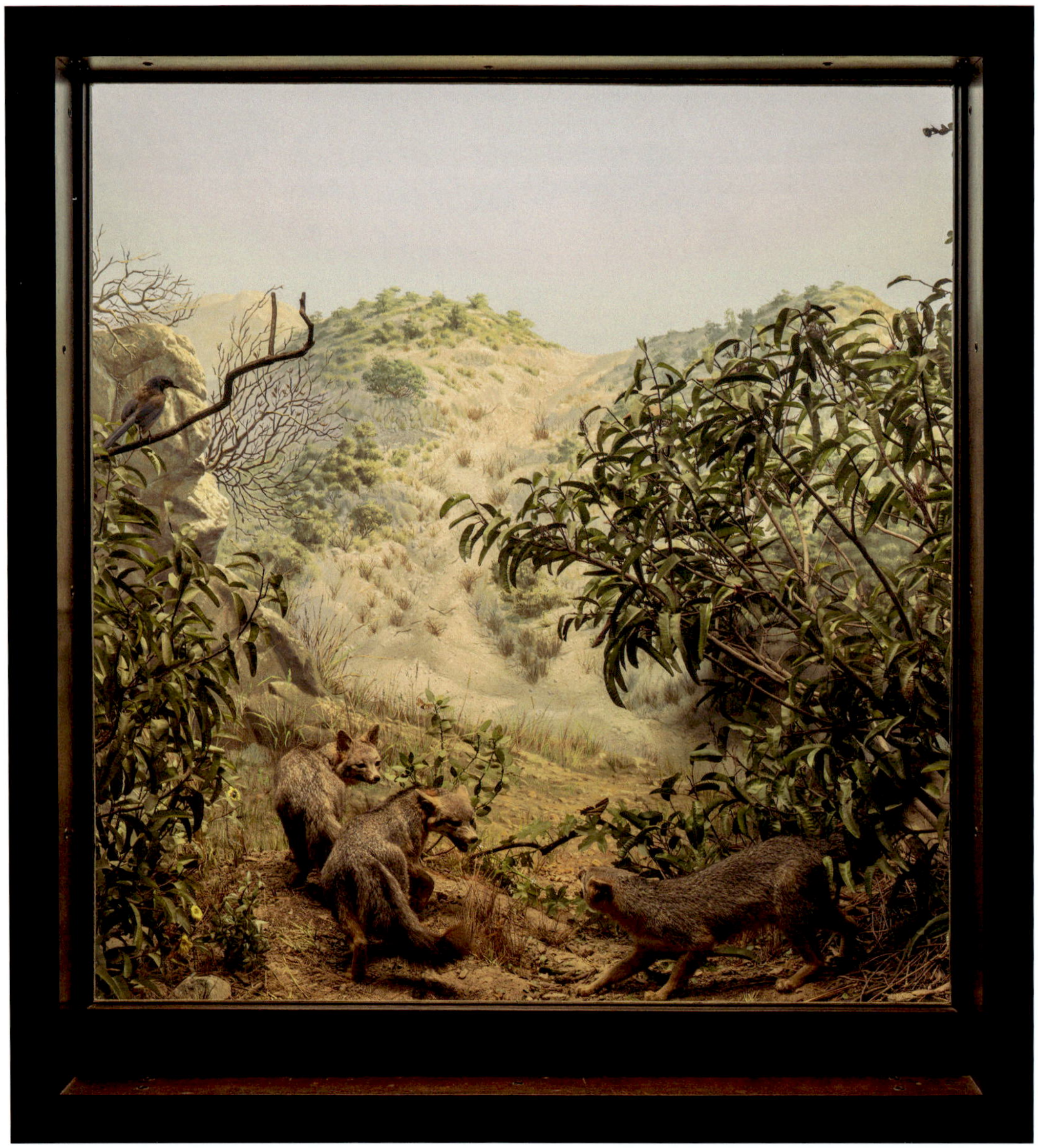

→ The sandy foreground of the *Gray Fox* diorama ties directly in with the background, creating one seamless view.

→ The *Pacific Marten* diorama uses a steep drop off a cliff as its tie-in, creating the illusion of great distance between the foreground and background.

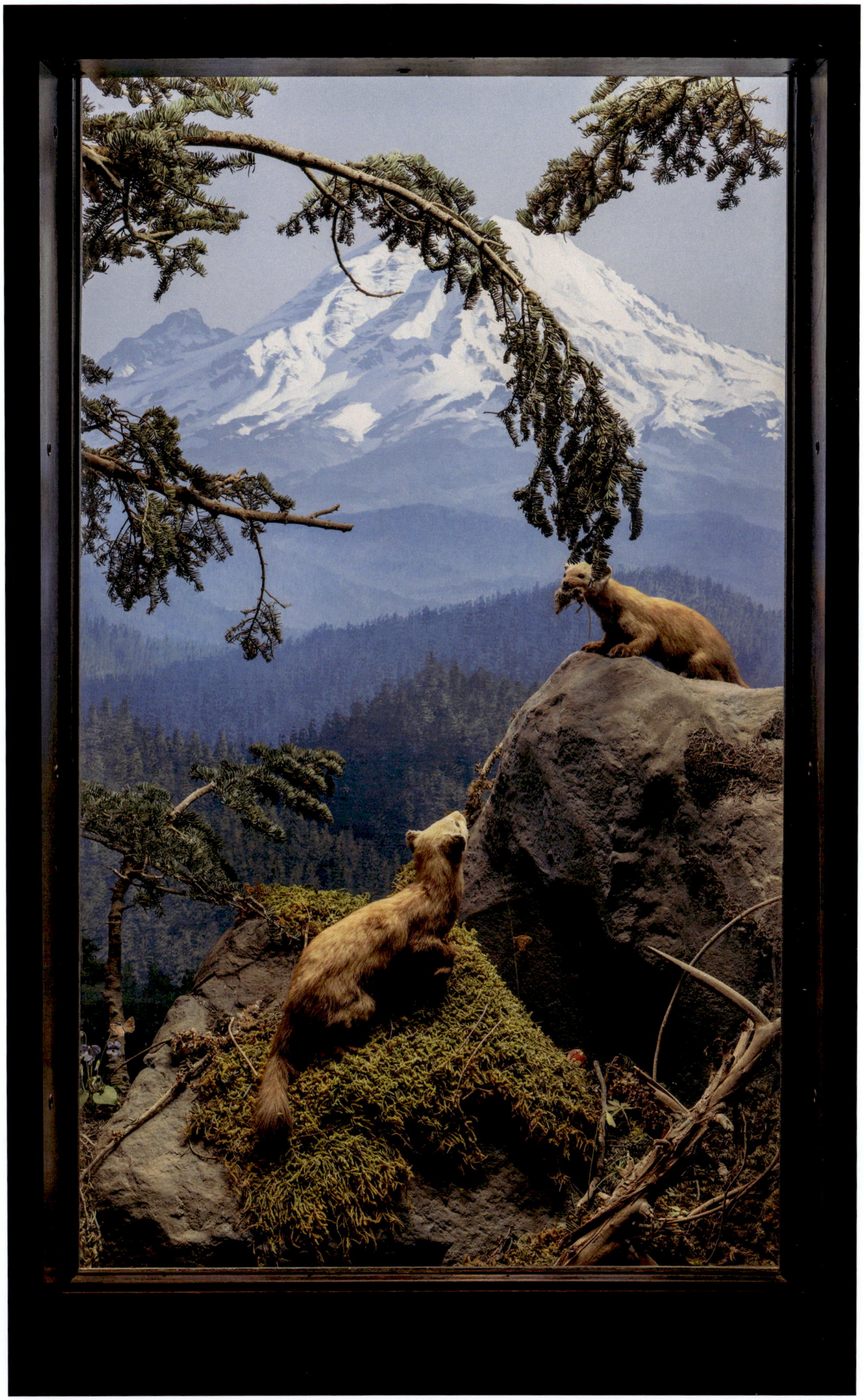

↓ Foreground Artist Peggy Clark paints a flower model for the *Eland* group. Although they are completely fake, plant models often look more realistic than dried plants, which can lose their vibrant colors during preservation (c. 1959).

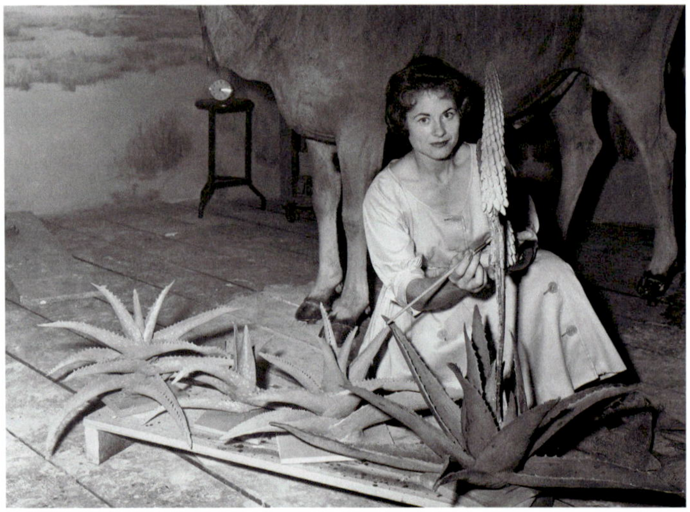

→ Realistic plant models are very difficult and time consuming to produce, so early dioramas often used real dried plants instead. Artist Bob Sewell sprays a living tree with latex to preserve it for inclusion in the *Gray Wolf* diorama (1951).

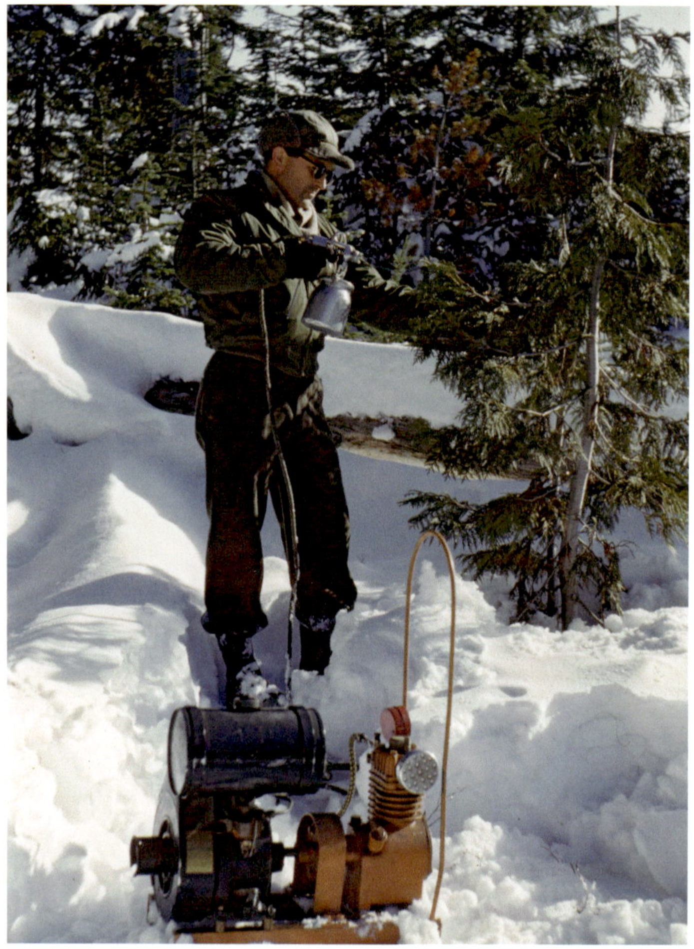

→ Foreground artists make detailed field studies of plants to help them construct more accurate plant models. This 1920s study of African plants includes notes on similar-looking plants found around Los Angeles that could be used as close substitutes (2023).

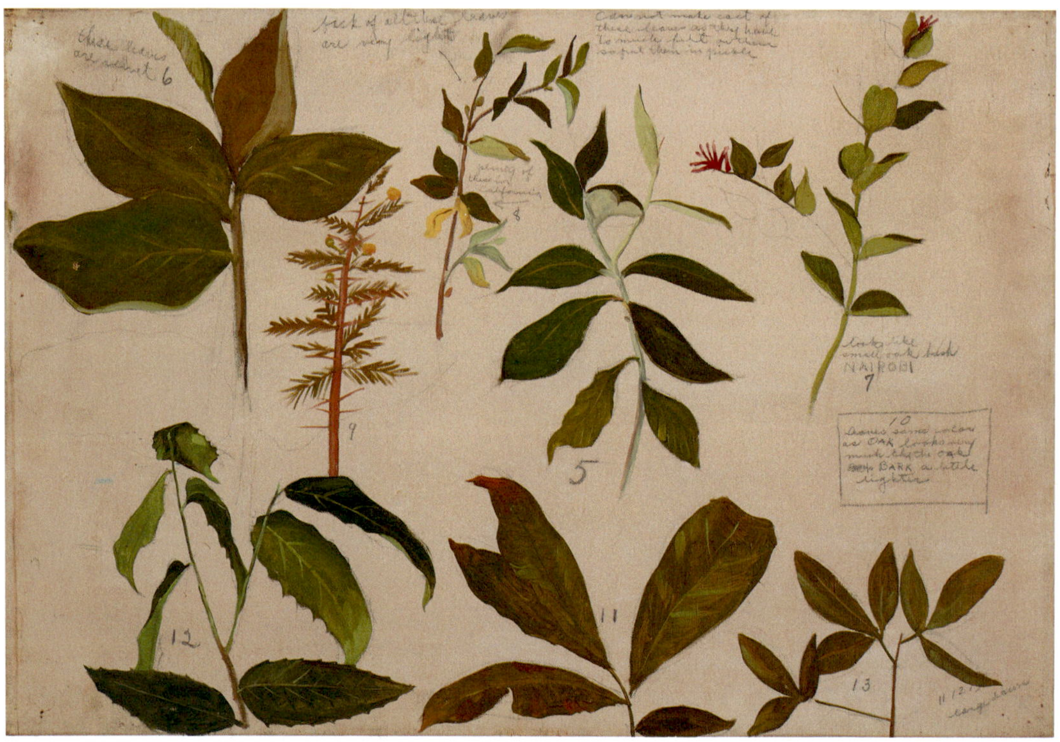

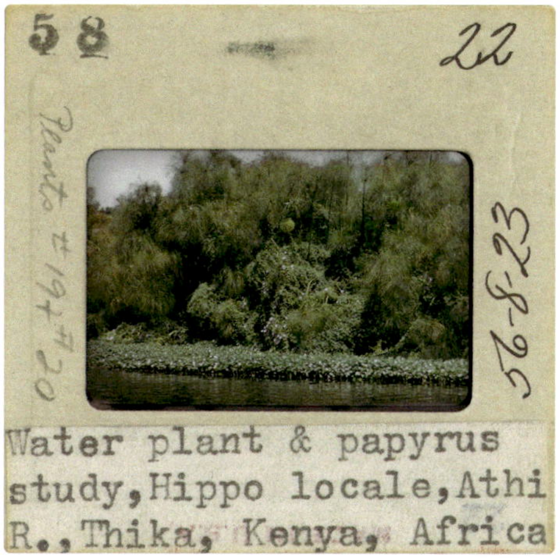

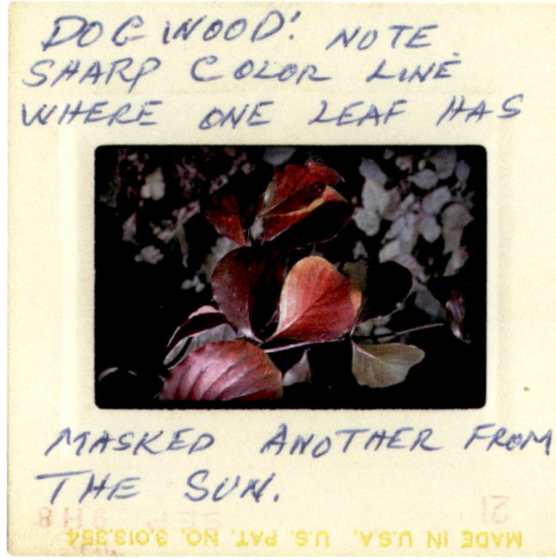

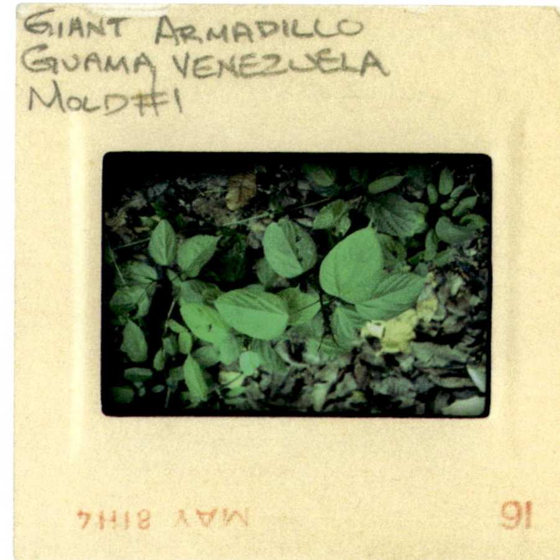

↑ Until the adoption of digital photography, background artists took most of their reference photos on colored slides like these three. The detailed notes on each slide linked the photos to specific molds brought back from the field (*left to right:* 1956, 1969, 1981).

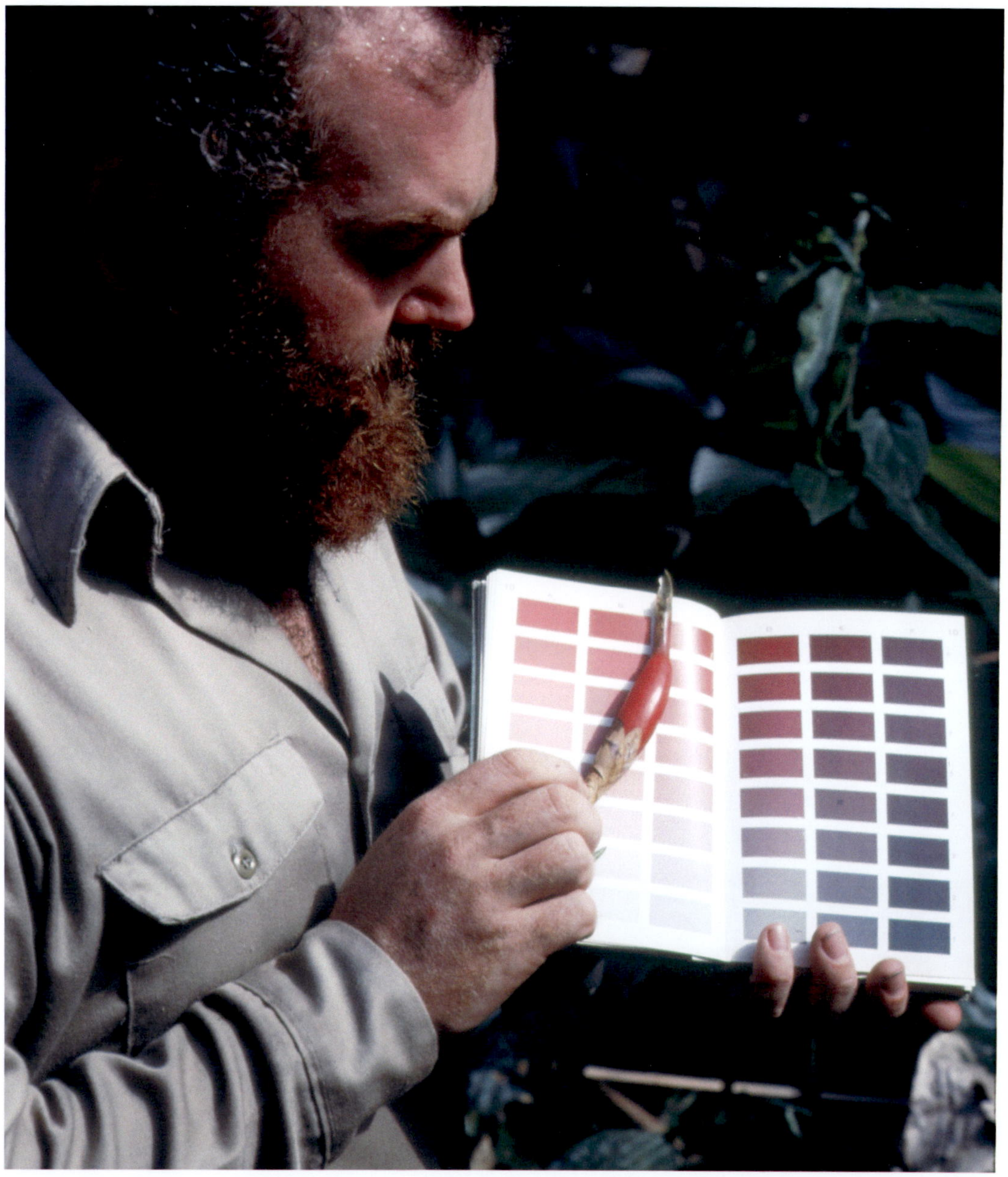

→ Chief of Habitats James Olson uses a color reference book to note down the exact hue of a red fruit found in Central Africa during the 1981 Logan African Expedition. Such color manuals are more reliable than photographs; they helped artists back at the museum reproduce the fruit's color accurately.

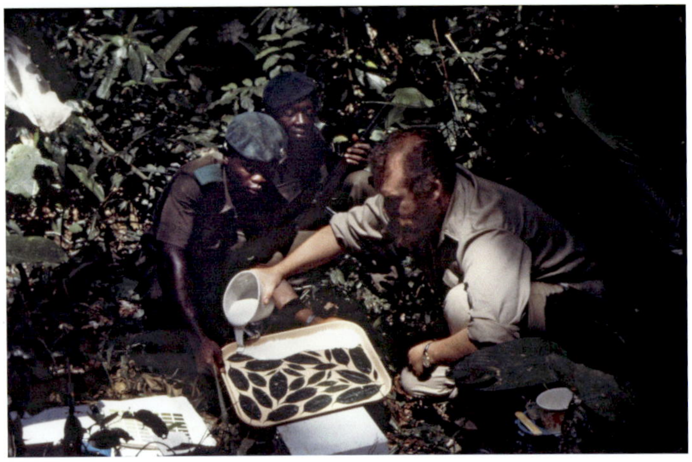

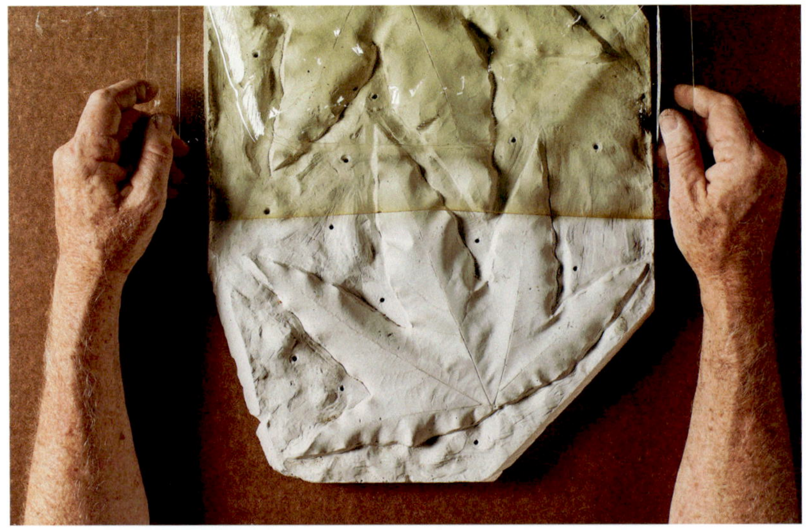

↑ Two park rangers in the Ituri Forest, Democratic Republic of the Congo, help Olson pour silicone over a tray of leaves to create a flexible mold. Back at the museum, the silicone mold was to be used to create a more durable plaster mold for vacuum forming replica leaves in the *Okapi* diorama (1981).

↓ Margaret Kincher demonstrates the laborious process of making artificial wax leaves one by one. With her left hand, she places a wire "stem" and a piece of thin cloth between two parts of a plaster mold. She would then pour hot wax from the stove to her right into the mold, clamping it together until the wax cooled. Any excess wax got trimmed away before she added the leaf to the pile on her right (1930s).

→ Although still laborious, the modern process of vacuum forming artificial leaves is an improvement over earlier methods. A hot plastic sheet is sucked onto a plaster mold and imprinted with leaf shapes (*top right*). The plastic leaves are then sprayed with a light coat of paint before being cut out (*middle*). The individual leaves are attached to wire stems and painted to resemble actual leaves, such as these ones made for a model sycamore tree (*bottom right*). To make the leaves even more realistic, holes are burned into them to simulate insect bite marks (2023).

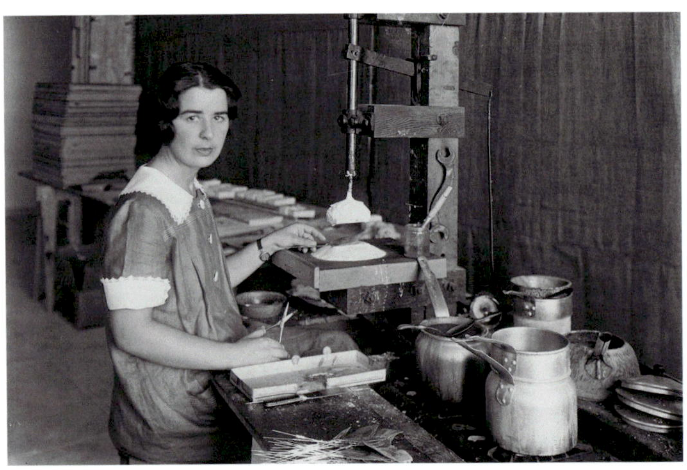

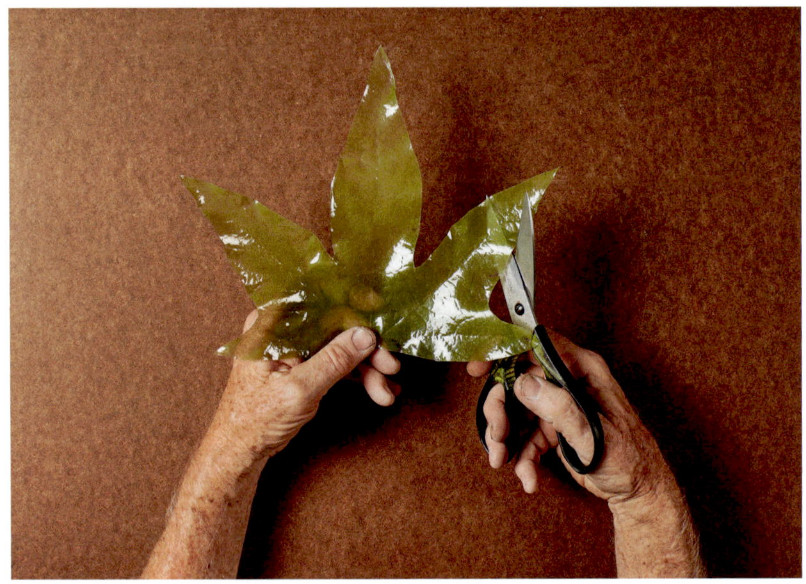

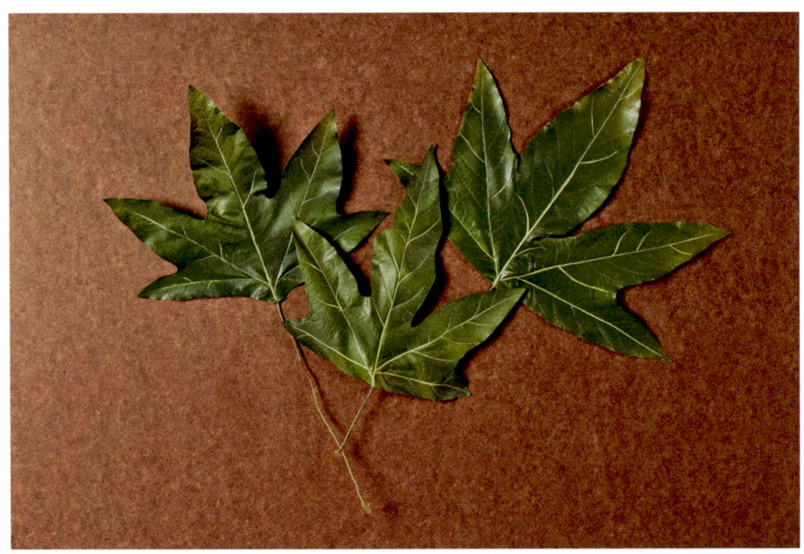

the actual location the diorama is portraying and make detailed studies of the flora there, photographing, sketching, and measuring all the plants they find so they can create accurate flowers and leaves back at the museum.

Various methods and materials have been used to produce leaves over the 100 years of diorama production at the museum, but most involve molding and casting. A mold is like an ice cube tray; the casts are the ice cubes themselves. A common way to make a mold is to pour liquid plaster over a leaf. When the plaster hardens, it reveals the leaf's detailed impression. Casting is when material like hot wax is then poured into that plaster mold. When the wax cools, it is pulled out to reveal the cast, a wax duplicate of the original leaf that the plaster mold was made from.

Early methods for producing replica leaves, such as wax casting, were quite time consuming, so dried plant materials were often used instead. Once the technique of vacuum forming over plaster leaf molds became possible in the 1960s, leaf fabrication on a large scale became easier. Vacuum forming starts out with a plaster mold of real leaves. If the diorama's location is close to the museum, a plaster mold is made directly from leaves in the field. If the location is remote, a rubber mold is made because a fragile plaster mold might break during shipping. The rubber mold is then used to make a plaster mold for vacuum forming at the museum.

The plaster mold with multiple leaf impressions is placed in a vacuum-forming machine with a sheet of clear plastic clamped over it that gets heated until it is soft. Then the mold and the soft plastic are brought together, and a vacuum sucks the plastic onto the mold, creating the impressions. The plastic cools in seconds and can be sent off to the next stage in the process.

A light coat of paint is sprayed on the plastic sheet so the leaves are easier to see before they are cut out by hand. Volunteers have helped greatly with cutting thousands of leaves out over many years. After the leaves

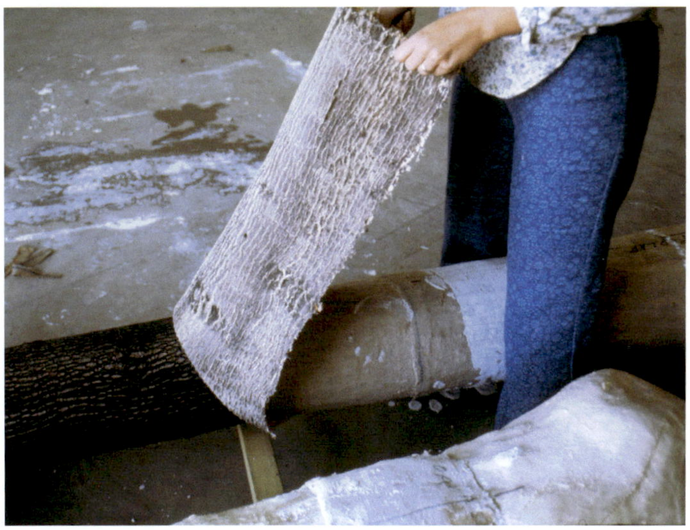

↑ A latex rubber mold is pulled off an actual tulip poplar tree. This mold will be pressed against plaster to create tree bark texture for a replica tree (1969).

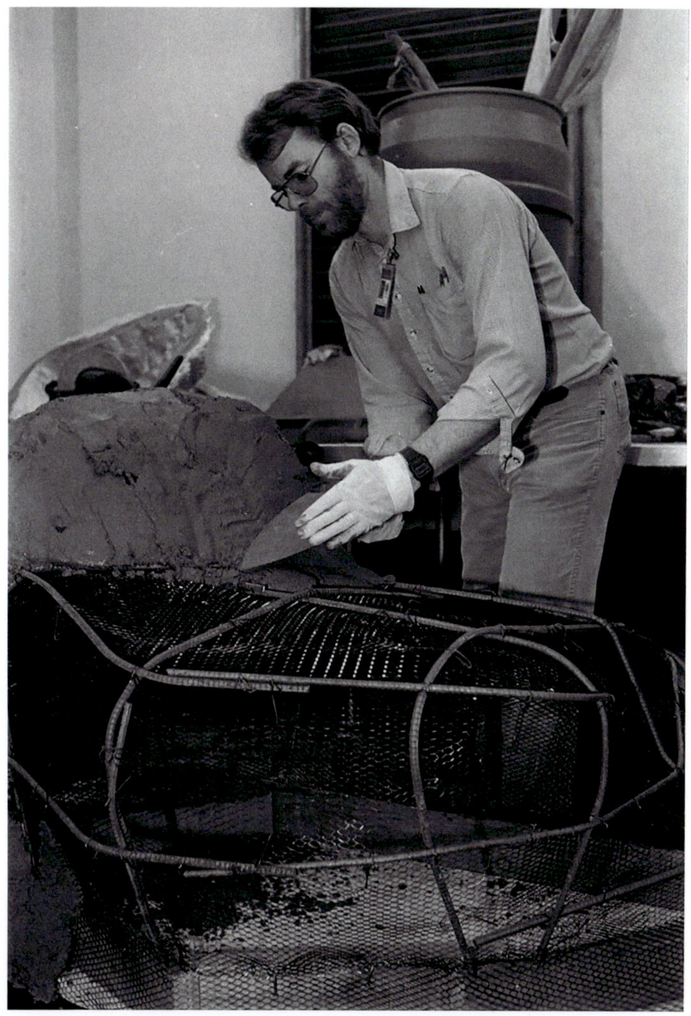

→ Taxidermist and Foreground Artist Tim Bovard sculpts artificial boulders for the *Kodiak Bear* habitat using concrete and rebar (1986).

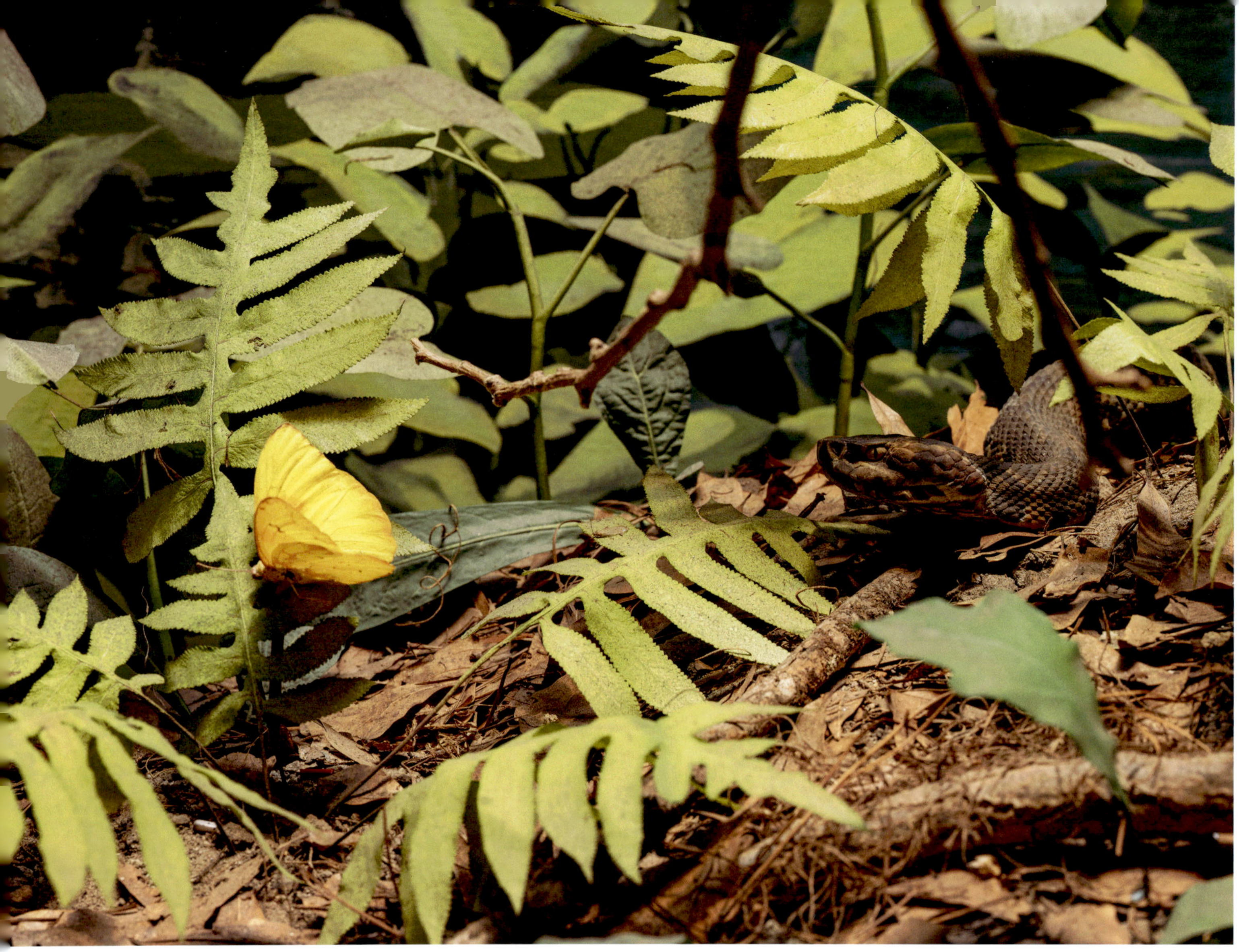

are cut out, a wire is glued onto the leaf to re-create the stem. Plastic tubing can be placed over the wire if the leaf needs a thicker stem. The leaf is then painted by spraying with a touch-up gun or airbrush. Multiple layers of color and even hand-painted veins work together to create the appearance of a realistic leaf. Some leaves are wilted or given replica bug bite marks to make the plants look even more realistic. Depending on the plant or tree, the leaves will be attached to natural woody branches and stalks or to artificial stalks and branches.

A single diorama may contain more than 10,000 hand-constructed leaves. As each leaf is handled a number of times through production, this process takes considerable time. That is why volunteers have been so important for leaf production at the museum. For example, plant production for the *Okapi* jungle habitat in 1985 required 72 volunteers and three staff working for three months.

↑ Often overlooked by visitors gravitating toward large taxidermied animals and grand background vistas, diorama foregrounds hold numerous hidden delights, like this water moccasin snake sneaking up on an orange-barred sulphur butterfly in the *North American River Otter* group (2021).

↗ The *American Black Bear* diorama's richly detailed foreground is made from a combination of real dried oak leaves and branches and artificial plant and fungi models that show the vibrant spring flora of Yosemite Valley (2021).

Large foreground elements like rocks or trees are usually fabricated right in the diorama bay during installation. The basic shape of the tree is usually made with wood and hardware cloth (wire mesh), or sometimes large, cardboard tubes. Tree bark is formed either with burlap

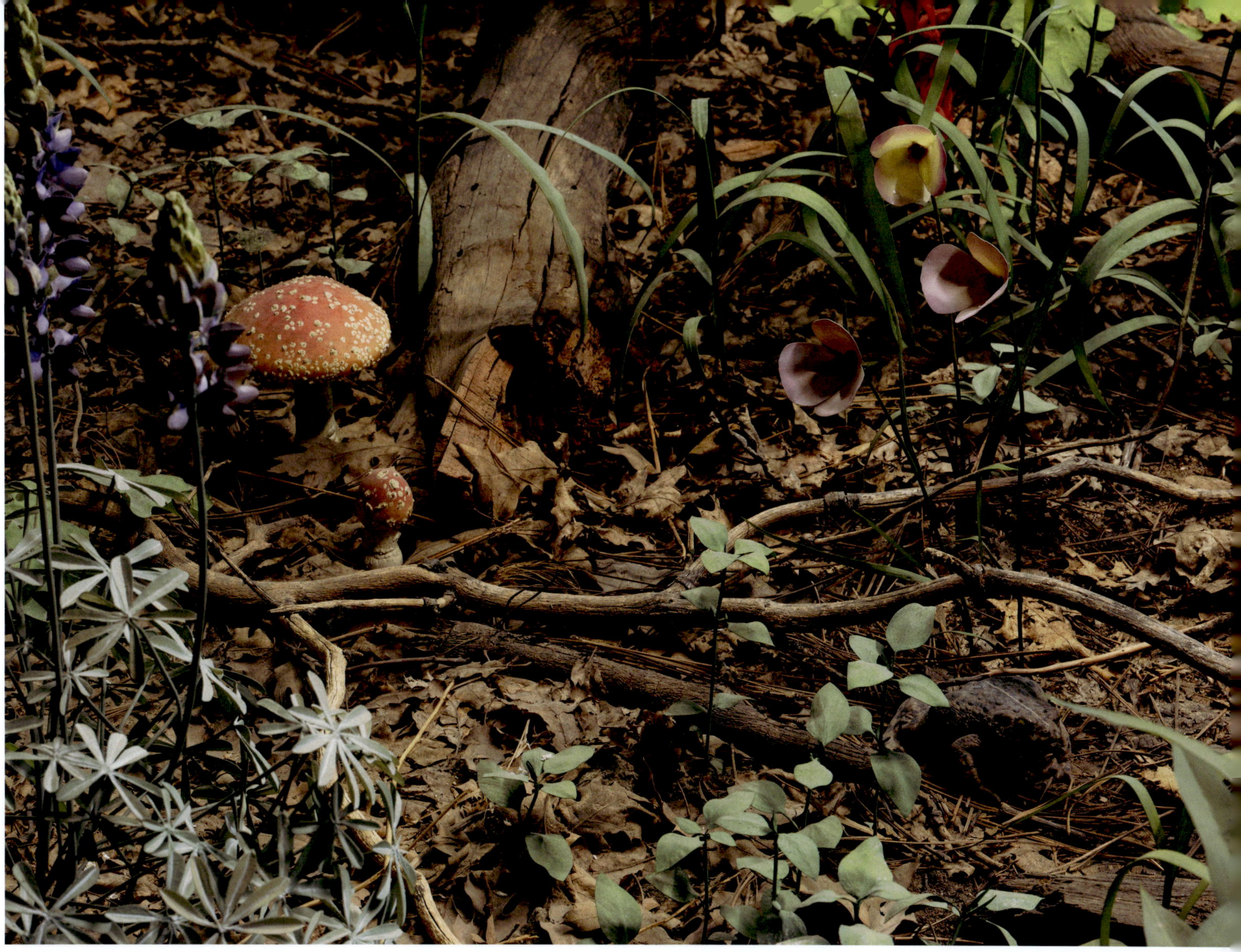

and plaster or with fiberglass and epoxy. Realistic bark texture is applied by first making a rubber mold of a trunk in the field and then pressing that mold into the plaster or epoxy. Finally, the tree is painted and leaves and smaller branches are added.

Real boulders would be too heavy to put in the dioramas, so large rocks are often made with wood and hardware cloth, then coated with burlap and plaster before a final texture is added. Sturdier rocks that visitors might sit on are made with rebar and concrete, which creates the illusion and feel of a real rock. Some rocks are very distinctive looking and need to be re-created exactly in the diorama; these are done using rubber molds of actual rocks in the field.

The illusion of water is particularly difficult to re-create in a diorama. Frozen water like snow is made by sifting ground-up plastic over cotton batting for a soft look. Ice is made from Styrofoam or clear plastic sheets and resin. Fluid water for streams or ponds is made from sheets of plastic or resin textured with little ripples or waves. Until major revisions of the halls in 2006, several NHM dioramas actually had real, flowing water in them!

The detail of the museum's foregrounds has definitely increased over time. Early foregrounds were often rocky and sparse, or else filled with easily obtained dried grass. But as new techniques were developed and skilled model makers joined the museum staff, foregrounds became increasingly detailed, leafy, and realistic. Foregrounds are often overlooked compared to the stunning vistas of background paintings, but they are amazing works of art in their own right and hide many small moments of discovery for those willing to look under a leaf or behind a rock.

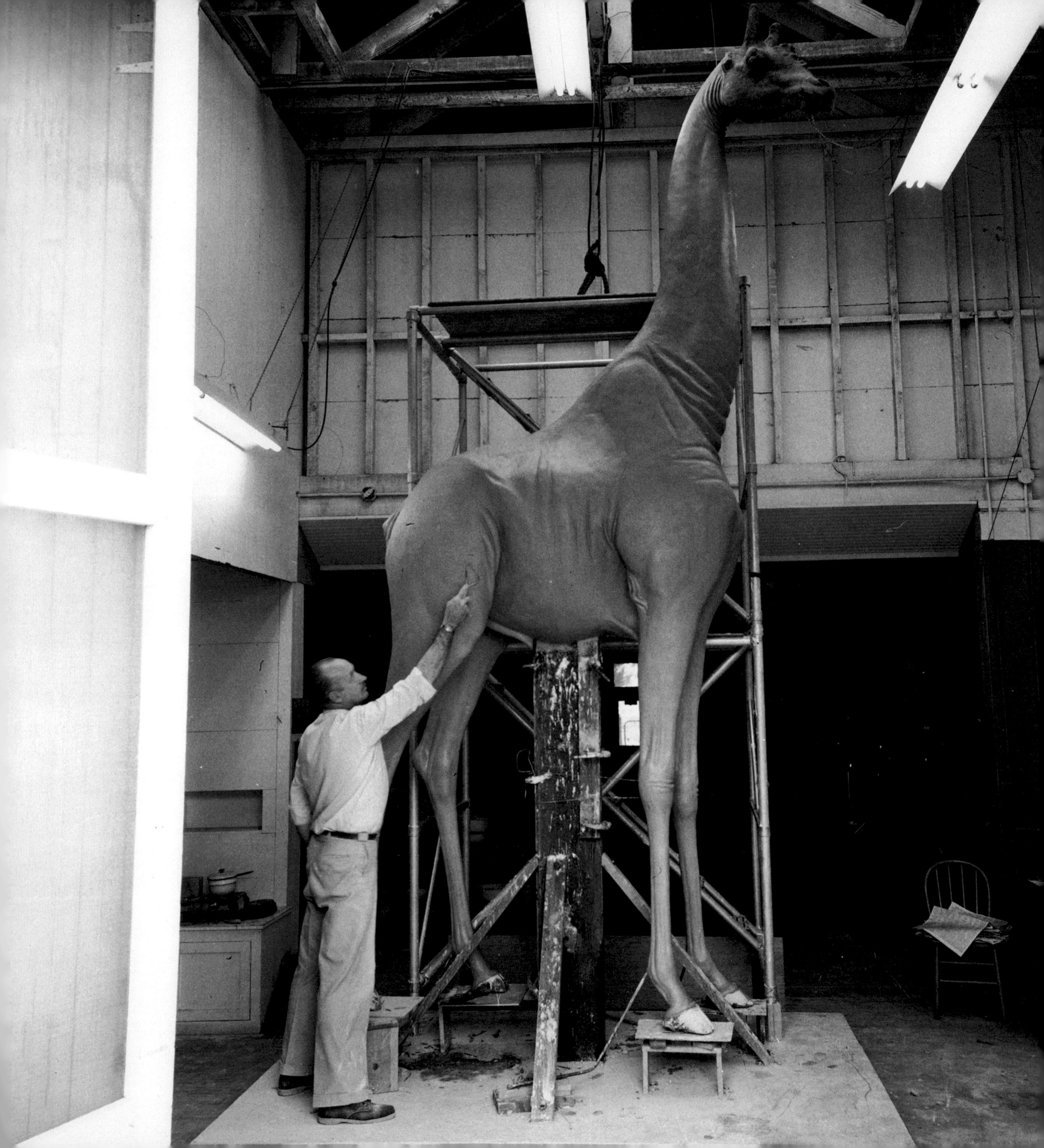

Taxidermy

Tim Bovard

← Taxidermist George Adams sculpts the underlying musculature of a giraffe in clay, demonstrating the incredible artistry required to produce lifelike mounts (1956).

Visitors to the Natural History Museum often ask if the animals in the dioramas are real. The answer is yes and no. The animals inside are taxidermy, a kind of art that combines detailed sculpture with real animal parts to create the appearance of a living animal. The fur, feathers, horns, and antlers that visitors see are real, but the eyes and the bodies underneath the preserved hides are re-creations. The fundamental challenge for a taxidermist is to take the real, preserved hide of an animal and transform it into an illusion of life. Taxidermy is the art of creating this illusion.

Taxidermy mounts are the focal point of the diorama; the main animal or animal group must capture the viewer's attention. The supporting characters, such as small mammals, birds, reptiles, amphibians, and insects, hide around the foreground, letting the viewer discover them. Each animal in the group must be posed to blend naturally into the final diorama scene.

Part of the overall decision to produce any diorama concerns which specimens the museum has in its collections and which it can acquire. For example, the museum may want to build an orangutan diorama, but to get a particular specimen can take many years. To get an endangered or rare species like an orangutan, the museum has to wait until one dies naturally. Because of this, NHM has often temporarily filled in bays with animals already in the collection, rather than keep dioramas empty waiting for the perfect specimen. Depending on the scene that the museum wants to create, multiple animals of different ages and sexes may be needed. In general, the museum tries to exhibit average, realistic specimens for any species, not the largest or most impressive animals that it can find.

Because NHM's dioramas open easily and are often changed and updated, they can feature a mix of specimens collected in different ways and different times. This mixture reflects a complex history of research needs, conservation regulations, and public sentiment. Prior to the 1980s, many specimens (except for those of rare and endangered species) were collected in the field. The taxidermist who accompanied the expedition ensured that the specimens were properly selected and prepared in the field for transport back to the museum and that reference materials were collected at the same time. Any collecting was, and still is, strictly regulated by government permits, and whenever possible expeditions collected animals that were already slated for removal. For example, the elephants in the current *Watering Hole* diorama

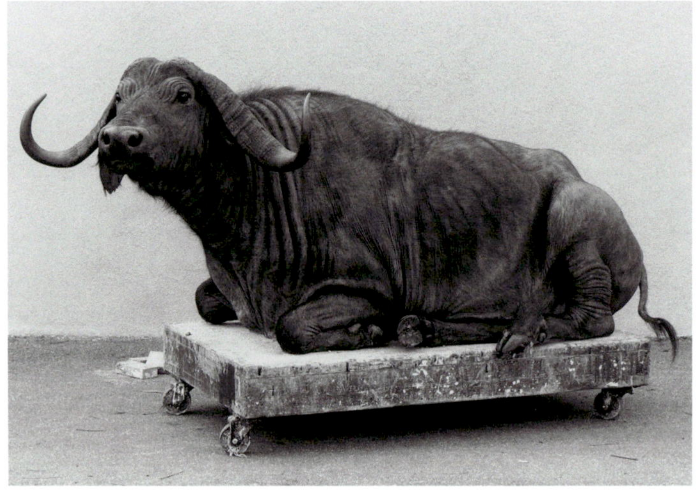

↑ What appears to be a Cape buffalo sitting down for a rest on a dolly is actually a taxidermy mount by George Adams waiting for transport to its diorama. Taxidermy uses the real skin of animals wrapped over a sculpted form to create a lifelike illusion (c. 1961).

↓ Taxidermist Tim Bovard in his NHM studio, whose back wall is covered in plaster death masks of various animals made by generations of taxidermists during the 100-year history of the museum's diorama program (2013).

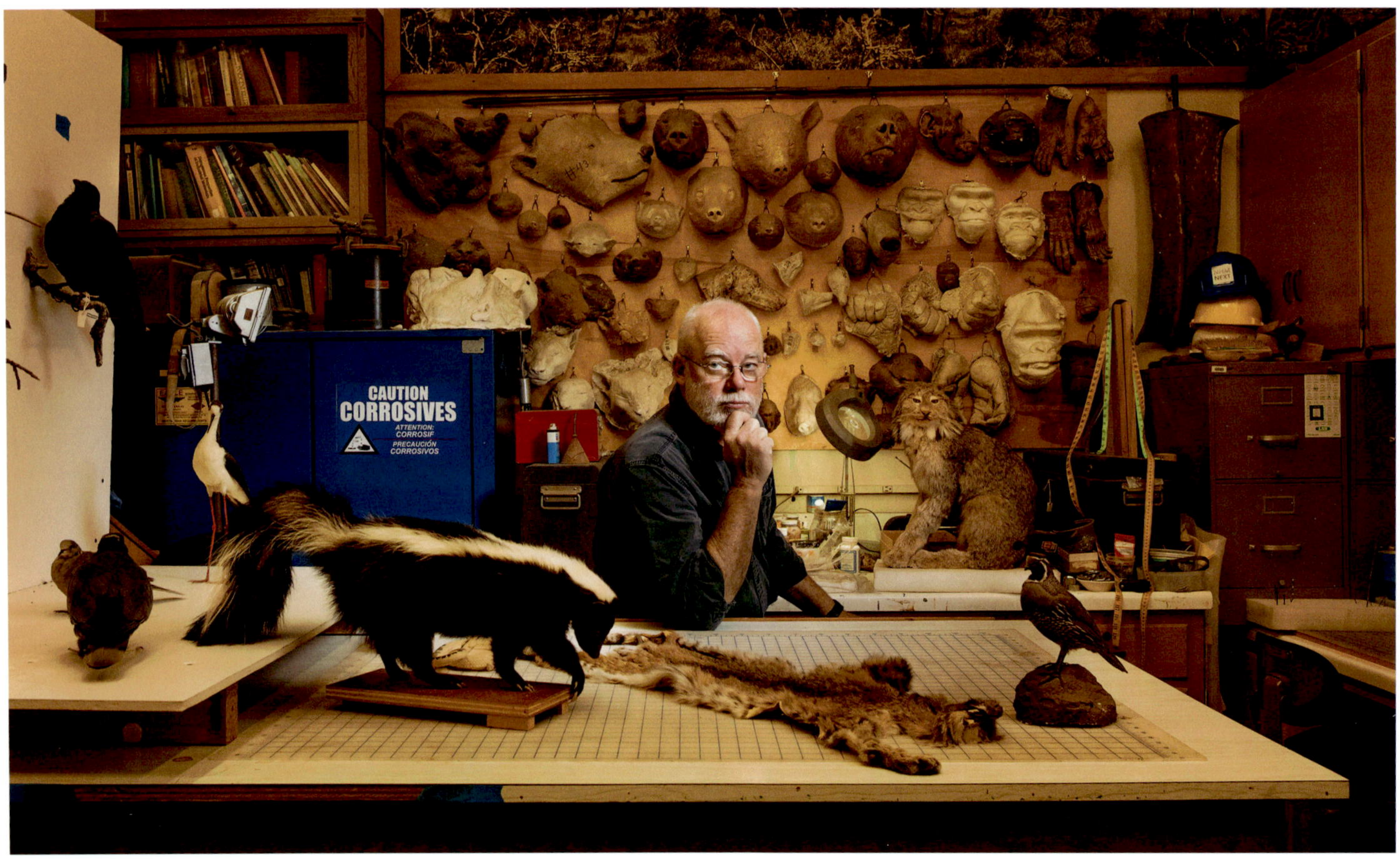

→ While the fur and feathers on taxidermy mounts are real, their eyes are glass replicas. This assortment belongs to a range of animals, from the pea-size eyes of a bird up to the quarter-size eyes of a wildebeest (2023).

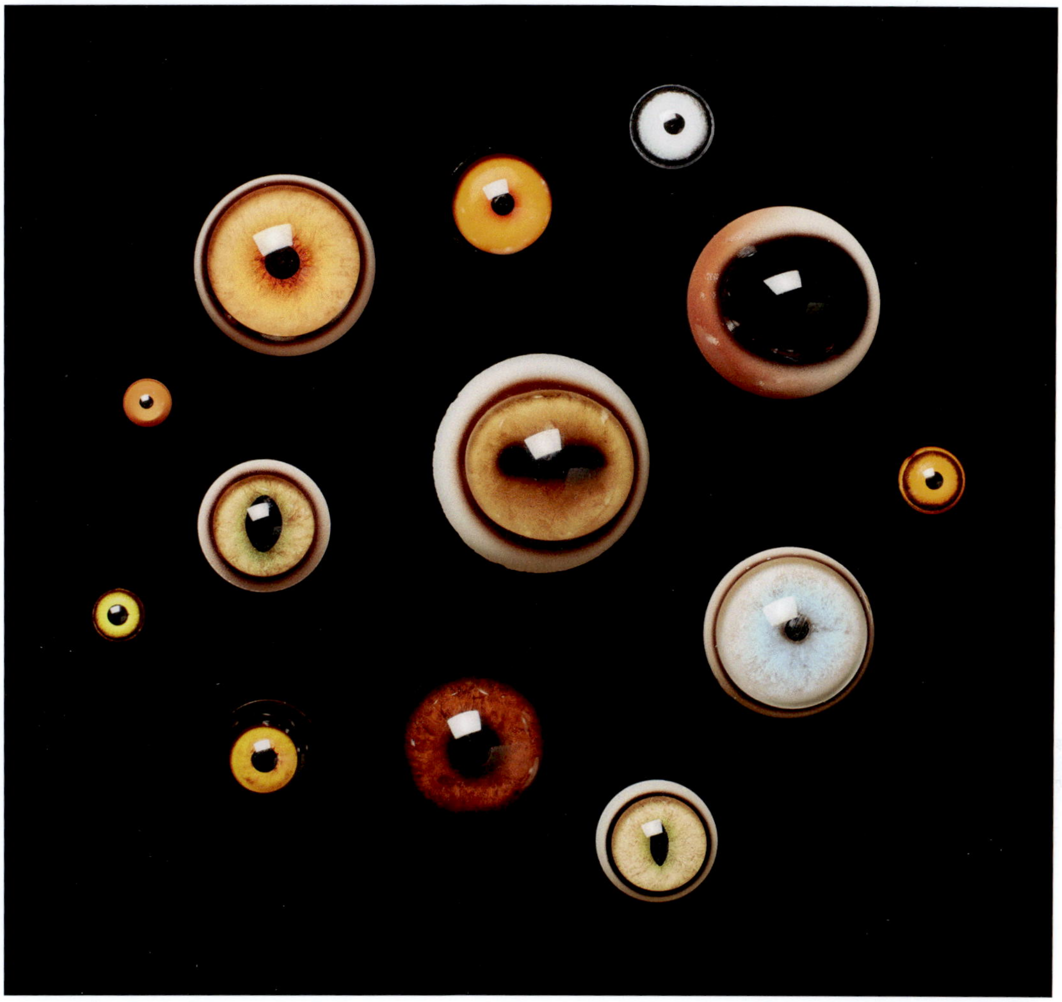

were just five of the nearly 4,000 elephants the Kenyan government planned to cull in the Tsavo area in the early 1960s to protect habitats and local farmers' fields from overgrazing. The animals in the *Walrus* diorama were collected by Yupik as part of their traditional subsistence hunting practices. The Indigenous hunters kept the walrus meat to eat but sold the valuable hides and skeletons to the museum to use.

Hunting large mammals used to be a fashionable pastime for wealthy museum donors and was enthusiastically supported by visitors, but over time, the public's view on hunting has grown more negative and the museum no longer needs to launch large expeditions to find specimens for habitat groups. Since the 1980s, most of the specimens prepared have not been collected in the field but have been salvaged instead. They may have been animals that died of natural causes in zoos or wildlife rehabilitation centers, birds that died of window strikes, road kill, or animals that wildlife agencies took as part of habitat management. Rare or endangered animals are always salvage specimens. When rhinos and elephants were collected for the dioramas over 60 years ago, they were much more common and legal to hunt, but the museum would never hunt these species today, because they are now endangered from heavy poaching.

Many different methods and materials can be used, but most mammal and bird taxidermy follows a similar overall process. The taxidermist first decides on a realistic pose for the animal that supports the narrative of the scene. Is the animal sleeping, running, fighting? An accurate manikin is created with this pose, and the skin is installed over it so that the animal looks alive. Any bare skin areas, like noses, eyelids, or inner ears, are recolored with paint after drying. Open mouths with teeth and tongues are usually painted models. Eyes don't preserve either, so glass eyes are used instead. Glass eyes have greatly improved in realism over the century of taxidermy at the museum. They give the viewer the sense that the

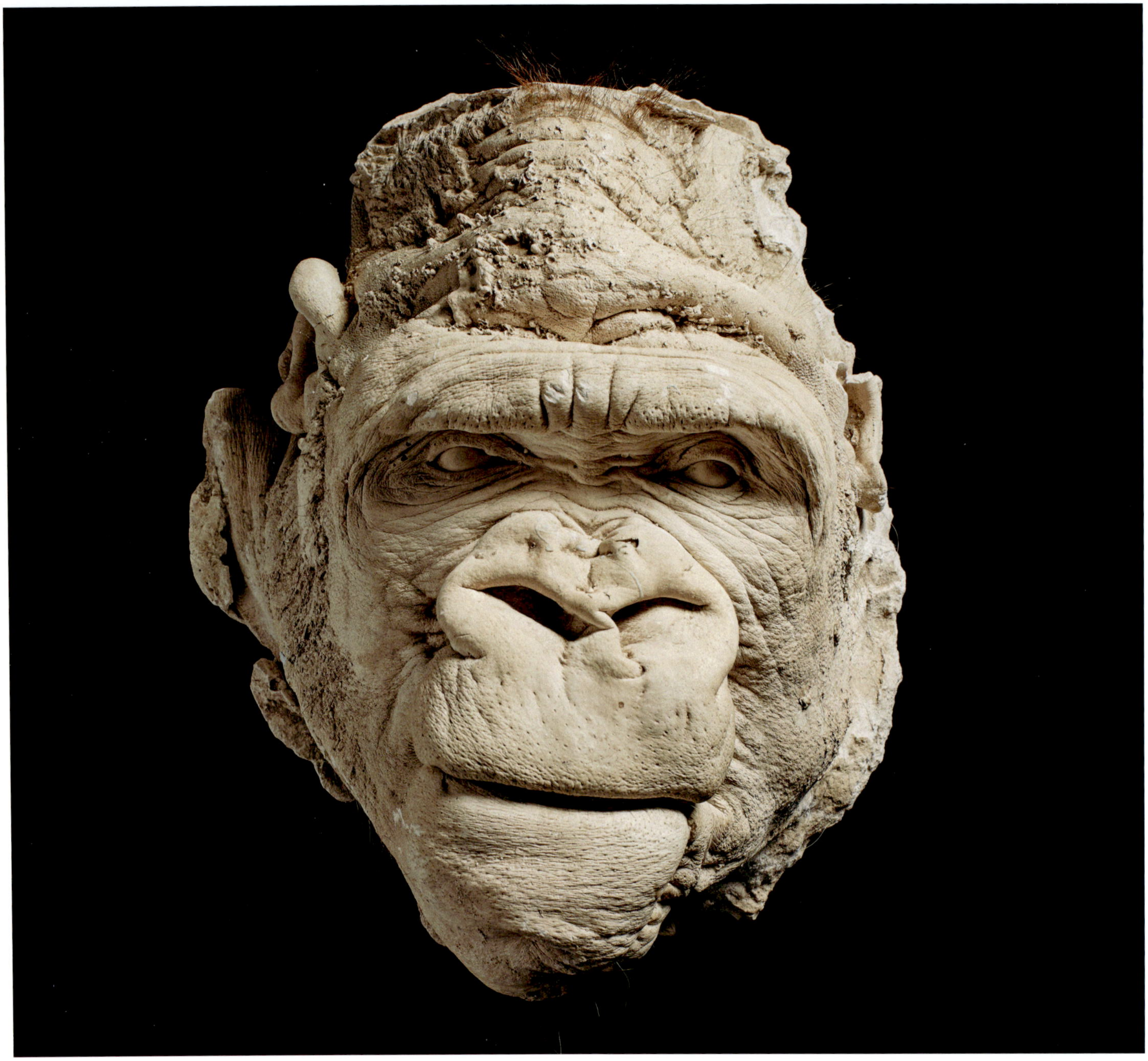

animal is actually there, living and breathing. Well-done taxidermy, with proper care, can make an animal look alive for more than a century.

It's crucial that a skilled taxidermist accompany any field expedition, as much of the work in creating a realistic mount happens in the field, before the hide ever reaches the museum. The process begins as soon as the animal dies. For mammals, the taxidermist takes detailed reference measurements all over the body before and after skinning it. Before skinning, taxidermists often make

↑ This plaster death mask was used to create an accurate replica of a gorilla's face that could be attached to the manikin and tanned hide to produce a realistic taxidermy mount (2023).

↓ With the aid of two field assistants, Taxidermist Adams uses a giant pair of calipers to measure features on an African elephant. The hundreds of detailed measurements later helped taxidermists sculpt an identically sized manikin in the studio that perfectly fit the elephant's hide (1963).

→ A member of the 1969 Logan Expedition prepares a thinhorn sheep's head to create a death mask that will accurately capture the animal's facial features.

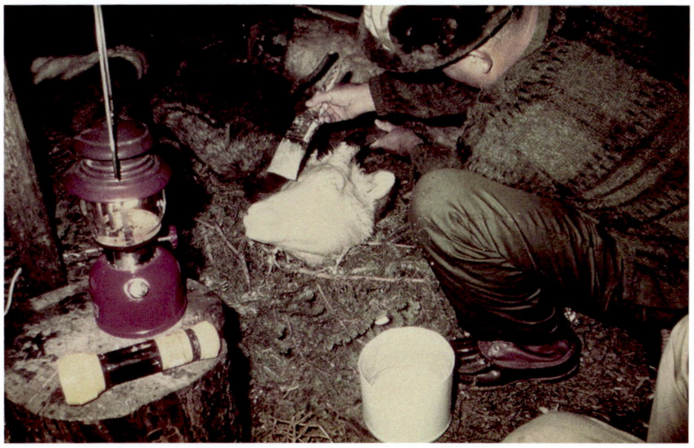

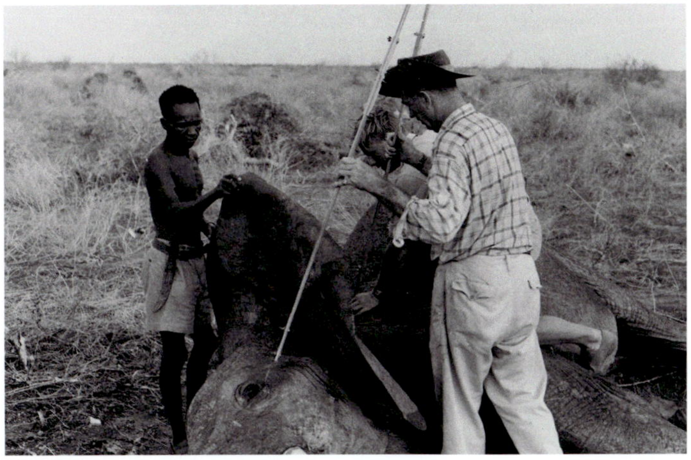

a death mask, a cast of the animal's face. After skinning, they may make a carcass cast of the whole body. These measurements and casts will later help in creating a realistic manikin that is exactly the same size as the original animal, so that the skin will fit over it properly. Once the skin is removed from the body, it is fleshed (scraped free of fat), the lips are split open, and the ears are turned (the ear skin is separated from ear cartilage). The skin is then salted multiple times and dried for transport. Back at the museum, the skin is rehydrated, pickled, tanned, rinsed, oiled, and dried out so that it can be safely stored until the taxidermist is ready to start making the mount.

Most of the mammal manikins that museum taxidermists produced prior to the 1980s were original sculptures. First, taxidermists would make an armature from the animal's real skeleton. Working from measurements, references, and photographs collected during field expeditions, they would then sculpt a clay model over the skeleton armature, accurately re-creating the muscles and form of the animal. To do this well, they had to be excellent sculptors and experts in the anatomy of hundreds of species. Once the sculpting was done, taxidermists built a plaster mold around the clay model. From that plaster mold, they would cast a burlap-and-plaster

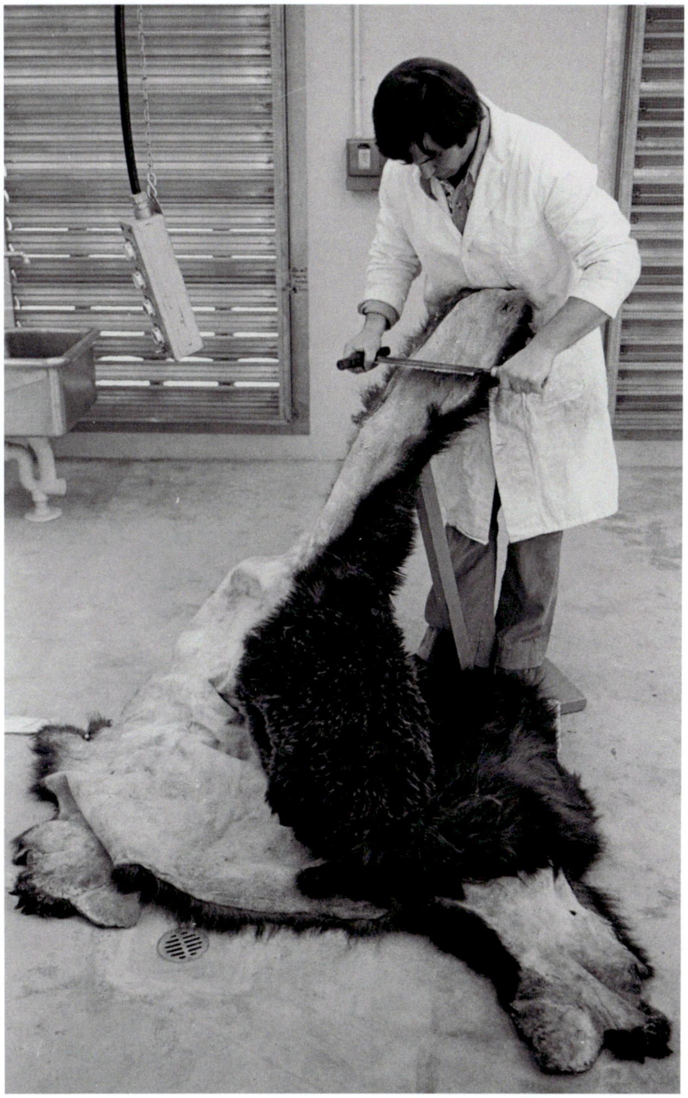

↑ A museum worker fleshes a hide to remove everything but a thin layer of skin (1970).

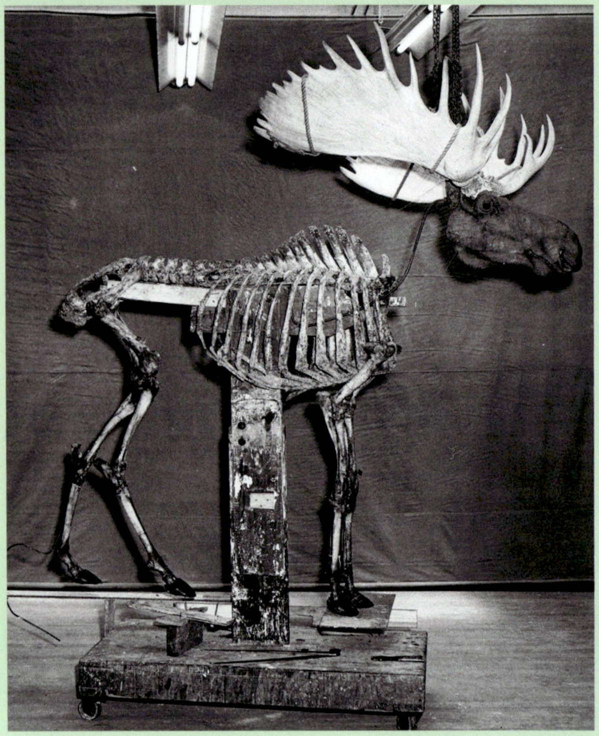

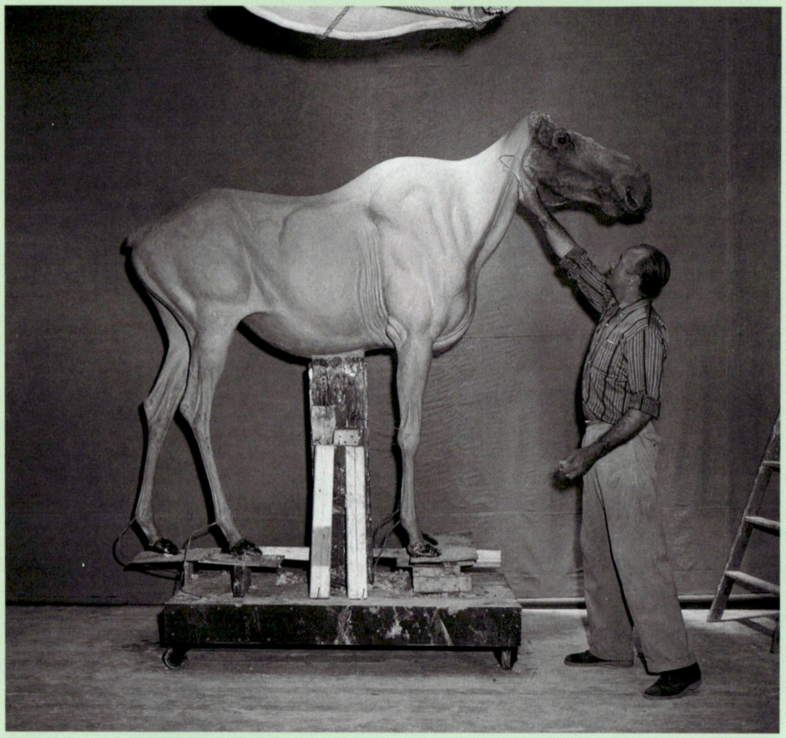

↑ Before the invention of foam manikins, taxidermy mounts started with an armature made from the animal's real skeleton, in this case a bull moose. The face is a replica based on the moose's plaster death mask (c. 1956).

↗ Using detailed field measurements, Adams uses clay to sculpt the muscles of the moose over its actual skeleton (c. 1956).

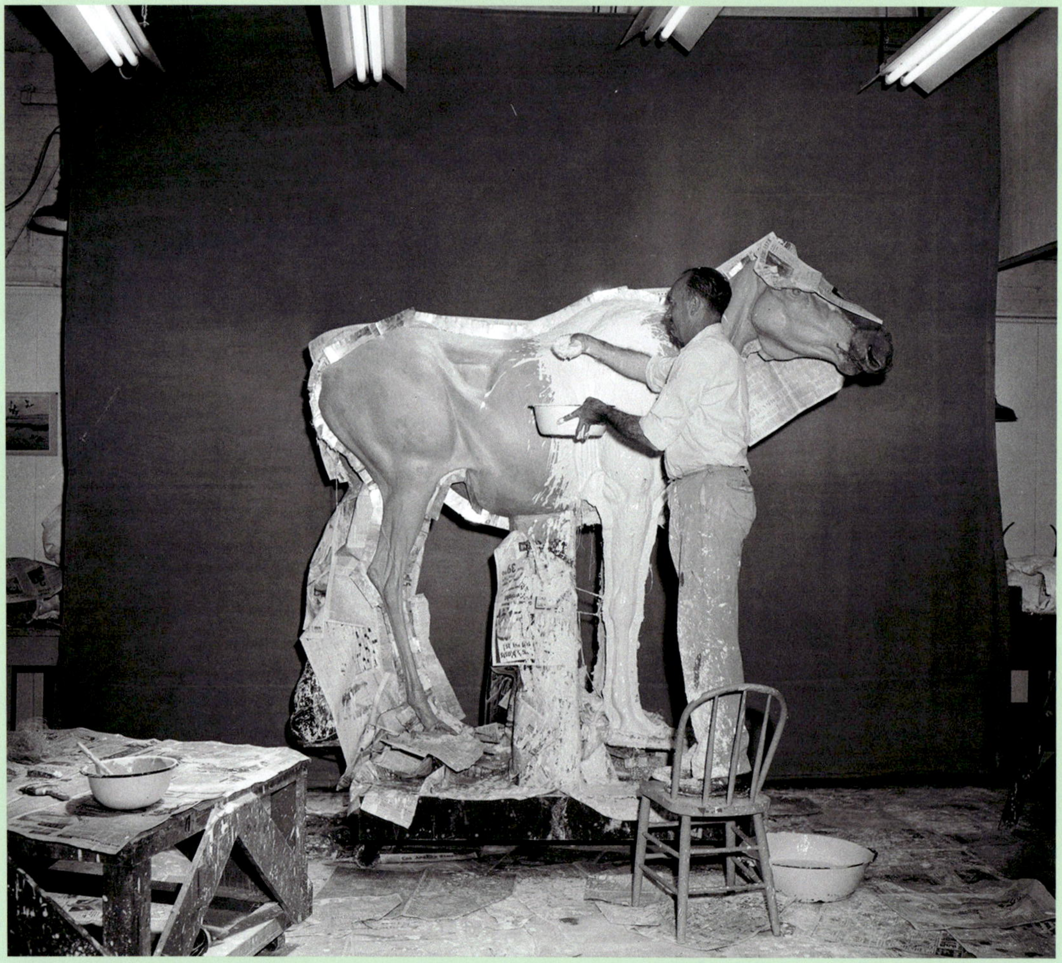

→ The finished clay sculpture is then covered in plaster to create a large mold (c. 1956).

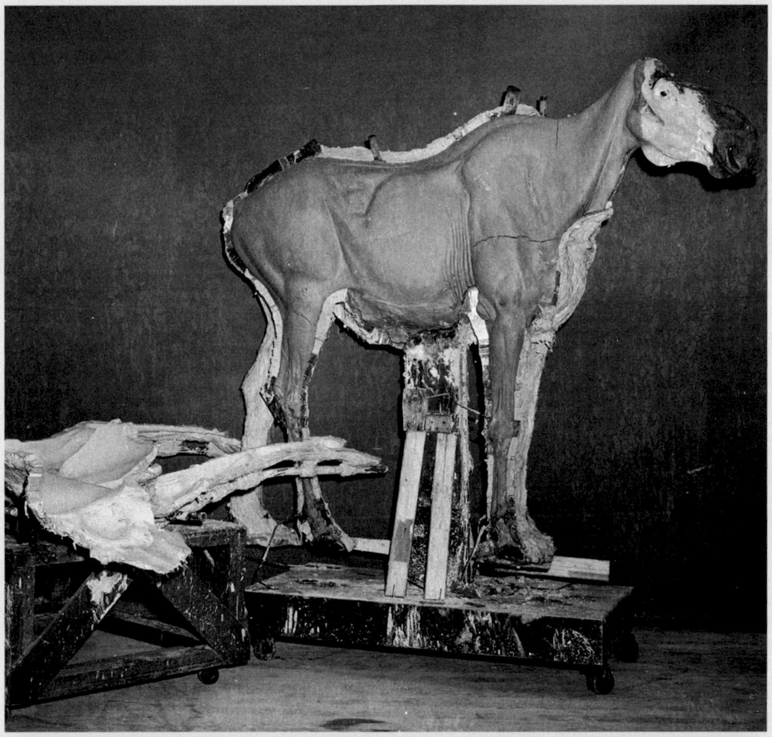

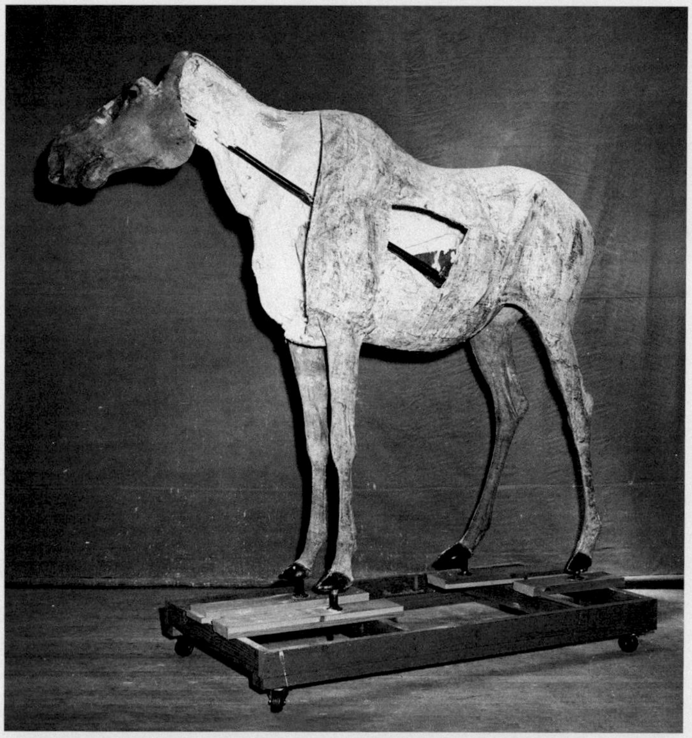

↖ The plaster mold removed from the clay moose sculpture will later be used to create the burlap-and-plaster manikin. Its purpose served, the clay sculpture is broken apart so that the animal's skeleton can be cleaned and placed into the museum's research collections (c. 1956).

↑ Burlap, soaked with plaster, is placed into the mold to create this strong but lightweight and hollow manikin (c. 1956).

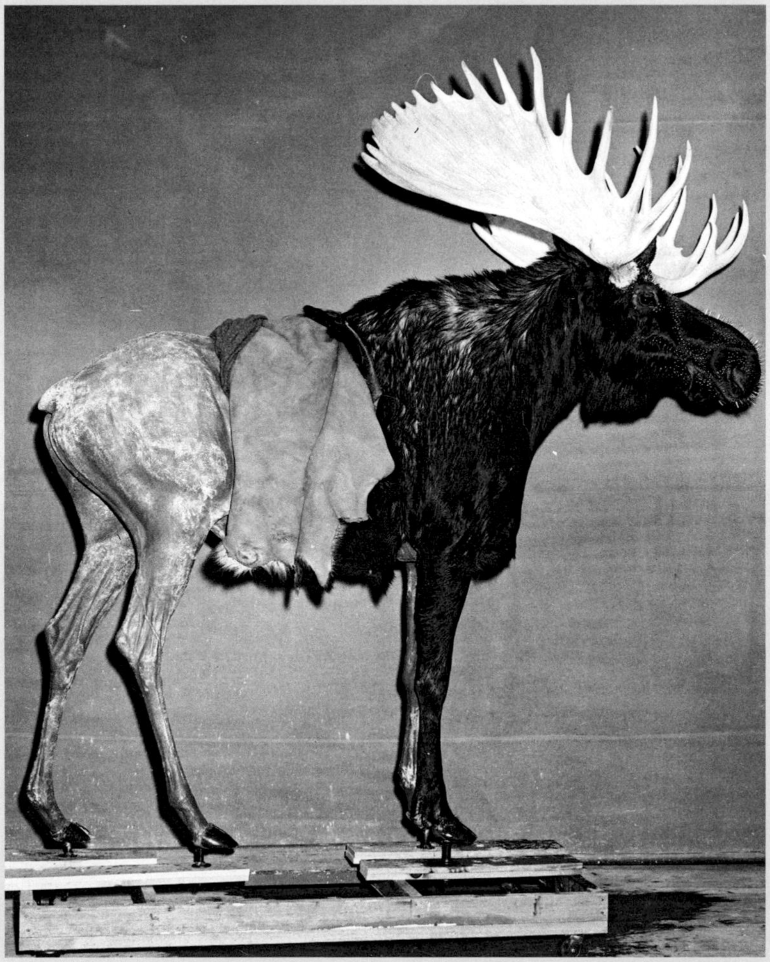

← The real hide of the moose is finally placed over the lightweight manikin and glued into place. The pins on the nose hold the skin in position while it dries. The real antlers are added back to the head and glass eyes are inserted to create what looks just like a living, breathing bull moose (c. 1956).

or papier-mâché manikin. This made a strong yet lightweight and hollow form that the skin could be wrapped around. The original clay sculpture was disassembled, and the skeleton was cleaned and placed in the museum research collections.

With the development of urethane foam in the 1980s, the manufacture of lightweight solid manikins became possible. Commercial taxidermy companies produce many manikins for common animals, which can be bought, resized, and resculpted to match a particular specimen, a process much faster than sculpting and molding a clay model to produce a manikin from scratch. Both methods can yield high-level results. The urethane forms are solid, unlike burlap-and-plaster or papier-mâché forms, which are hollow. They are, therefore, easily cut and reshaped for size and position. After resculpting, they can be foamed back together and installed with leg rods for mounting. Some forms may be cut into 30 or more pieces before being foamed back together. When no commercial manikin is available for an animal, the museum taxidermist will carve a manikin directly from foam or sculpt one in clay before making a fiberglass mold to produce the final foam manikin.

After the manikin is produced and the tanned skin has been rehydrated, that skin can be test-fitted over the manikin. If the skin fits well, the mounting process begins. The taxidermist inserts earliners that stretch the skin out into natural ear shapes. They add clay to the manikin for detail. For example, clay is used to set the glass eyes and shape the eyelids, ears, lips, nose, and feet. Clay can also be added in certain areas to form wrinkles. The skin is applied to the manikin with a slow-drying glue before

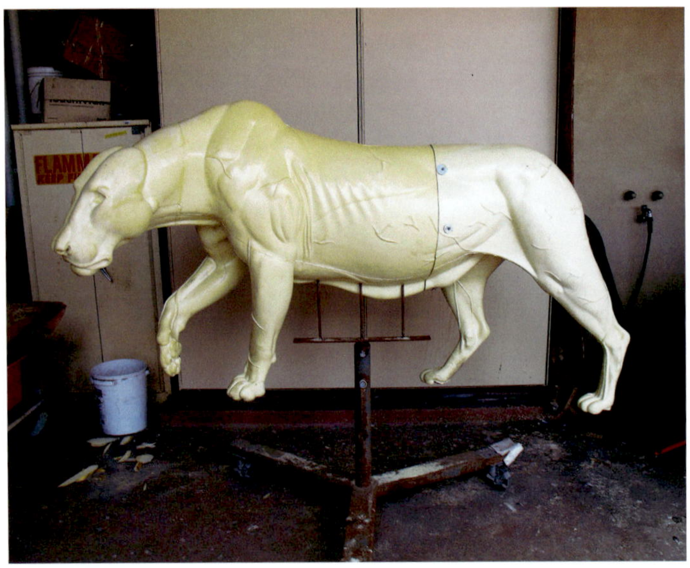

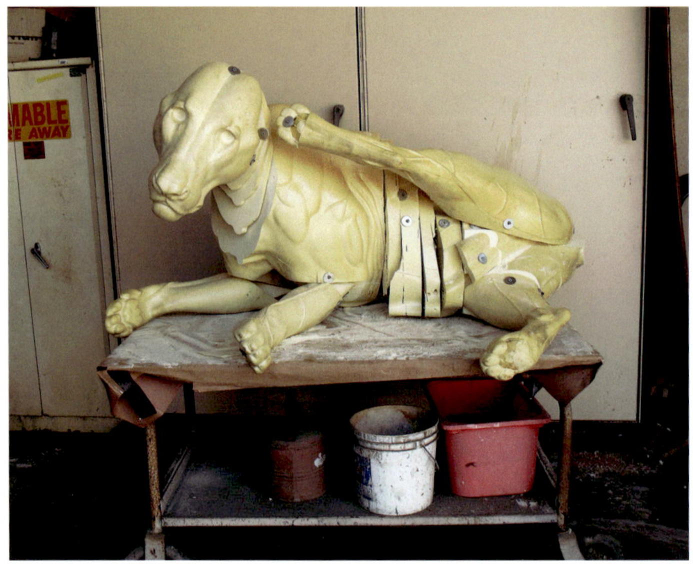

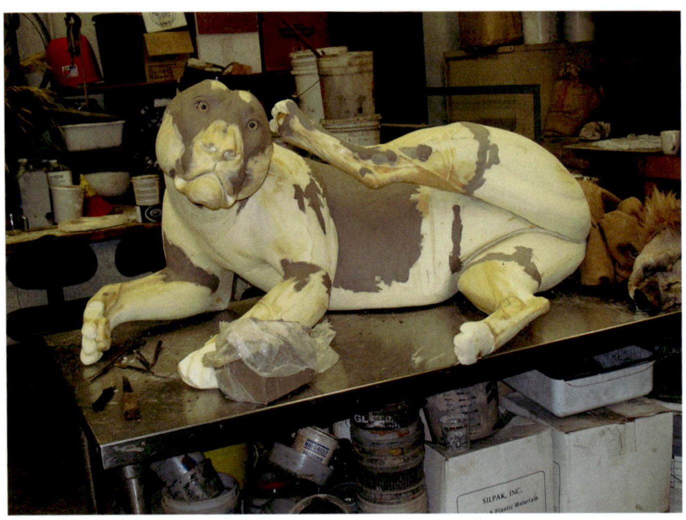

↖ Most modern taxidermy mounts start with commercially made foam manikins like this male lion in a walking pose (2019).

↑ The manikin for the walking male lion is cut up and posed as a lion scratching behind his ear (2019).

← With glass eyes added, rough edges smoothed over, and details resculpted with clay, the custom manikin is now ready to be wrapped in the lion's hide (2019).

→ Since John Rowley began the dioramas at NHM, museum taxidermists have avoided dramatic action poses in favor of capturing quiet, everyday moments like this striped skunk family foraging for grubs (2021).

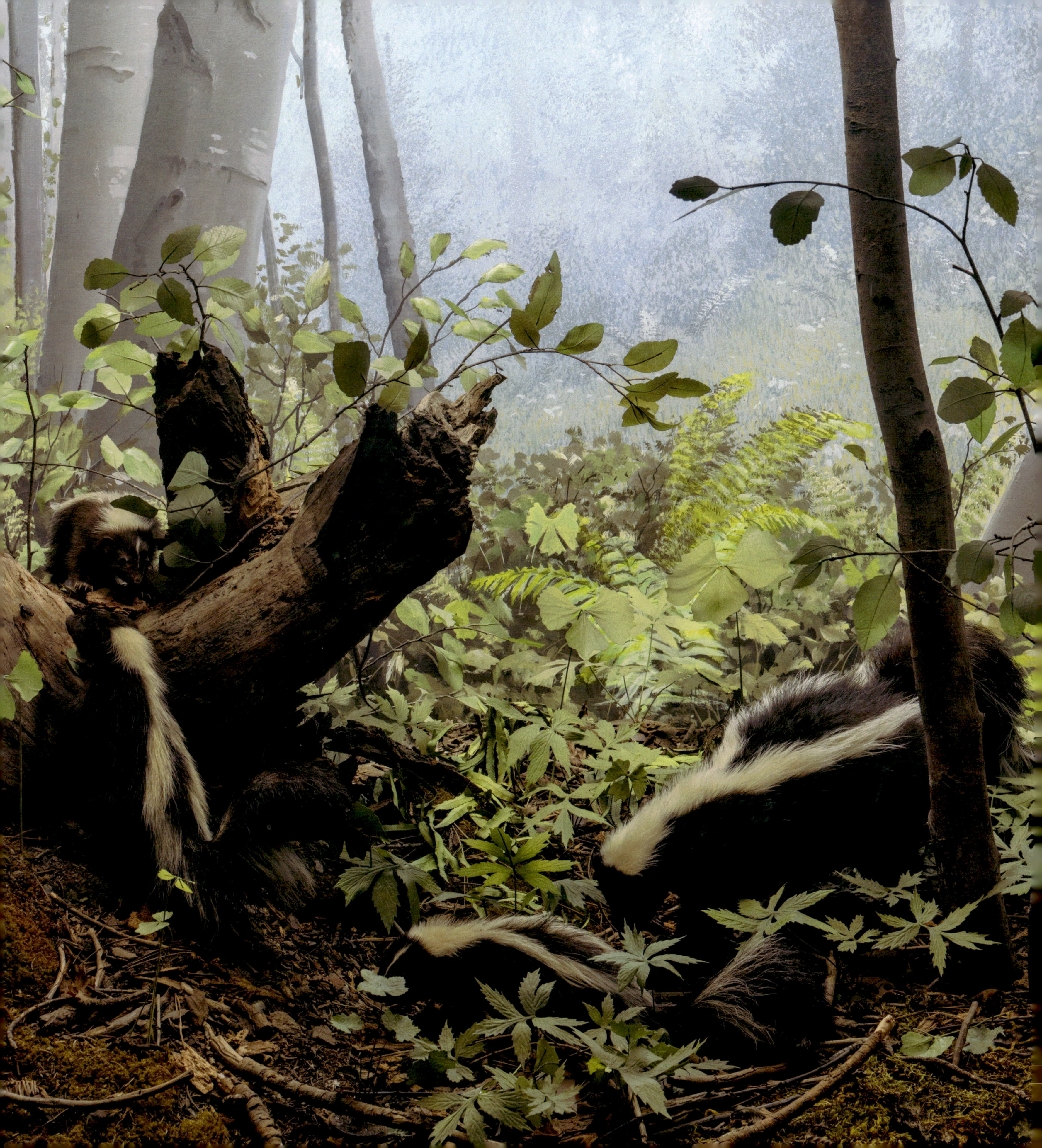

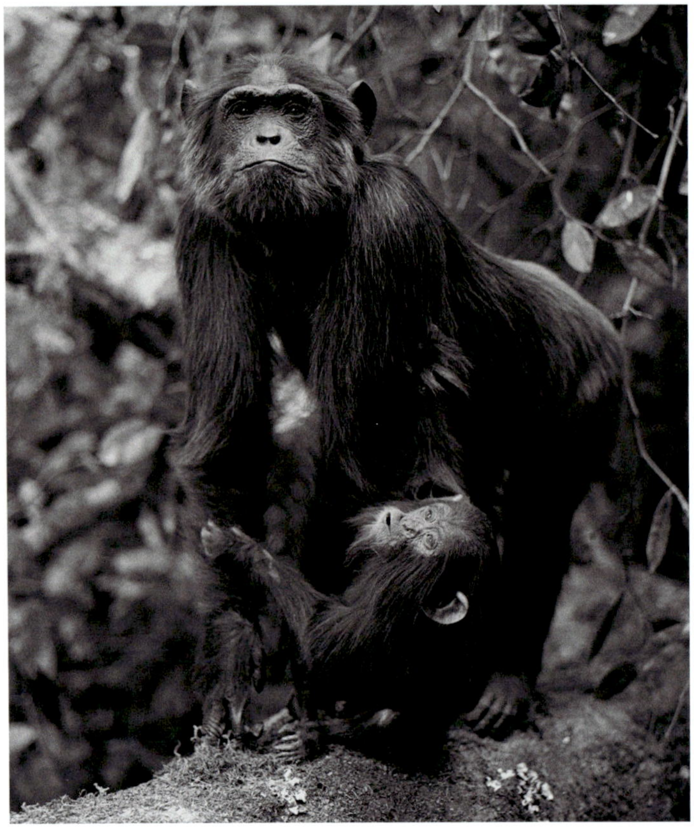

← It is difficult to accurately reproduce bare skin areas like the faces on these taxidermied chimpanzees. Adams used a complicated technique called wax infiltration to preserve the actual faces, hands, and feet of these animals instead of making models of them (1970).

↓ Unlike mammals, most bird manikins are made from the real skull, wing, and leg bones of the bird attached to a body and neck made of excelsior (wood shavings) wrapped in wool twine (2023).

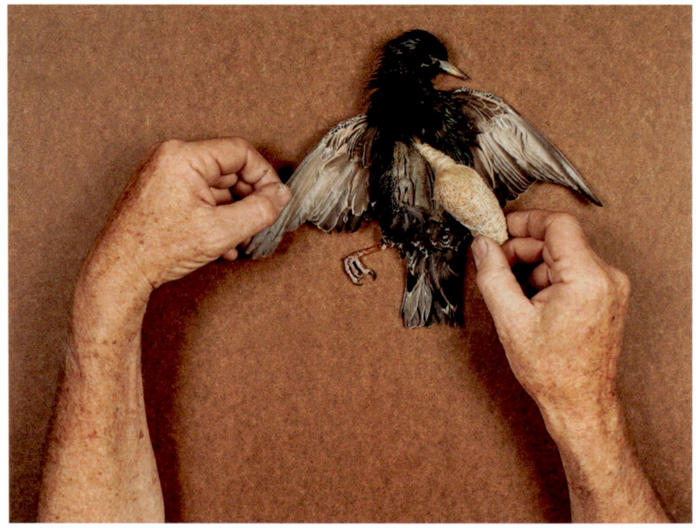

being sewn together around it. To give the mount a soft-flesh look, the taxidermist works to adjust the face, detail the feet, and add in wrinkles in the fur as the skin dries.

Certain mammals, like chimpanzees and gorillas with bare-skin faces, are more challenging to mount in a realistic way. In the 1960s, Taxidermist George Adams used wax infiltration to mount chimpanzees. This many-step, time-consuming process, which preserves the actual flesh by filling it with wax, can produce realistic bare skin areas such as faces, ears, hands, and feet. The method captured a more textured, realistic skin on the chimps than tanned, bare skin would have. These wax-infiltrated parts were attached to the manikin and the tanned skin for the final mount. For the gorilla mounted in the 1990s, the face and hands were detailed models based on molds of the gorilla's features taken soon after his body was donated by a zoo.

As with mammals, there are many methods and materials used to mount birds. Most birds are skinned with the first cut along their bellies. The legs, wing bones, and skull are left attached to the skin. A template tracing of the carcass that is made for reference purposes helps to produce an accurate manikin later. The taxidermist must remove all the flesh and fat manually from the skin before washing it in dish soap. Typically only large birds' skins are tanned like mammal hides. The skin is often treated with a bug poison before final mounting.

Most of the birds on exhibit at the museum were mounted on manikins made from excelsior (wood shavings) wrapped in string or thread. Some large birds are mounted on burlap-and-plaster or urethane foam manikins, as well. The bird's neck is made by a wire wrapped with string and inserted into the manikin. Taking the skin, the taxidermist inserts wires into the wings and legs, tying them to the bones still left there. Clay is added to the skull before glass eyes are roughly placed into position. The manikin is then placed into the skin with the neck wire inserted into the skull. The wing and leg wires are attached to the excelsior body, and the bird is sewn up.

The bird is then placed on a temporary base, and a blow dryer is used to dry and fluff the feathers. When the feathers are dry and the skin is still damp and flexible, the bird is positioned and preened for the final look. Unlike mammals, birds can be posed and positioned after mounting due to the flexible wire frame. As with mammals, any bare-skin areas must be recolored with paint.

Fish, reptiles, and amphibians can be taxidermy skin mounts over a form just as mammals or birds are, or even freeze-dried, but most commonly they are replicas. This is because the skin of fish, reptiles, and amphibians doesn't preserve in the same way as bird and mammal skin and must be painted instead. For replicas, the specimen is molded and a cast is made in fiberglass or plastic. Historically, plaster was used for the final replica. In all cases, whether it is real skin wrapped over a manikin or a replica, the specimen must be repainted to give it lifelike colors. Arthropods like insects and crabs are usually real specimens that have been positioned and dried out. Soft-bodied animals such as slugs have to be made from models sculpted from scratch.

For all museum taxidermy, whether slug or polar bear, the goal is to create the illusion of a live animal. The imaginative leap from a lifeless body or pelt to a lifelike mount is enormous. The process of making this leap requires artistic skill and the capacity to work with a wide range of materials and animals. A good taxidermist is an artist with experience in many areas, including collecting animals, skinning, tanning, sculpture, anatomy, molding and casting, modeling, painting, and design. NHM's continuing diorama program has allowed a new wave of taxidermists to learn this art and gain these unique experiences.

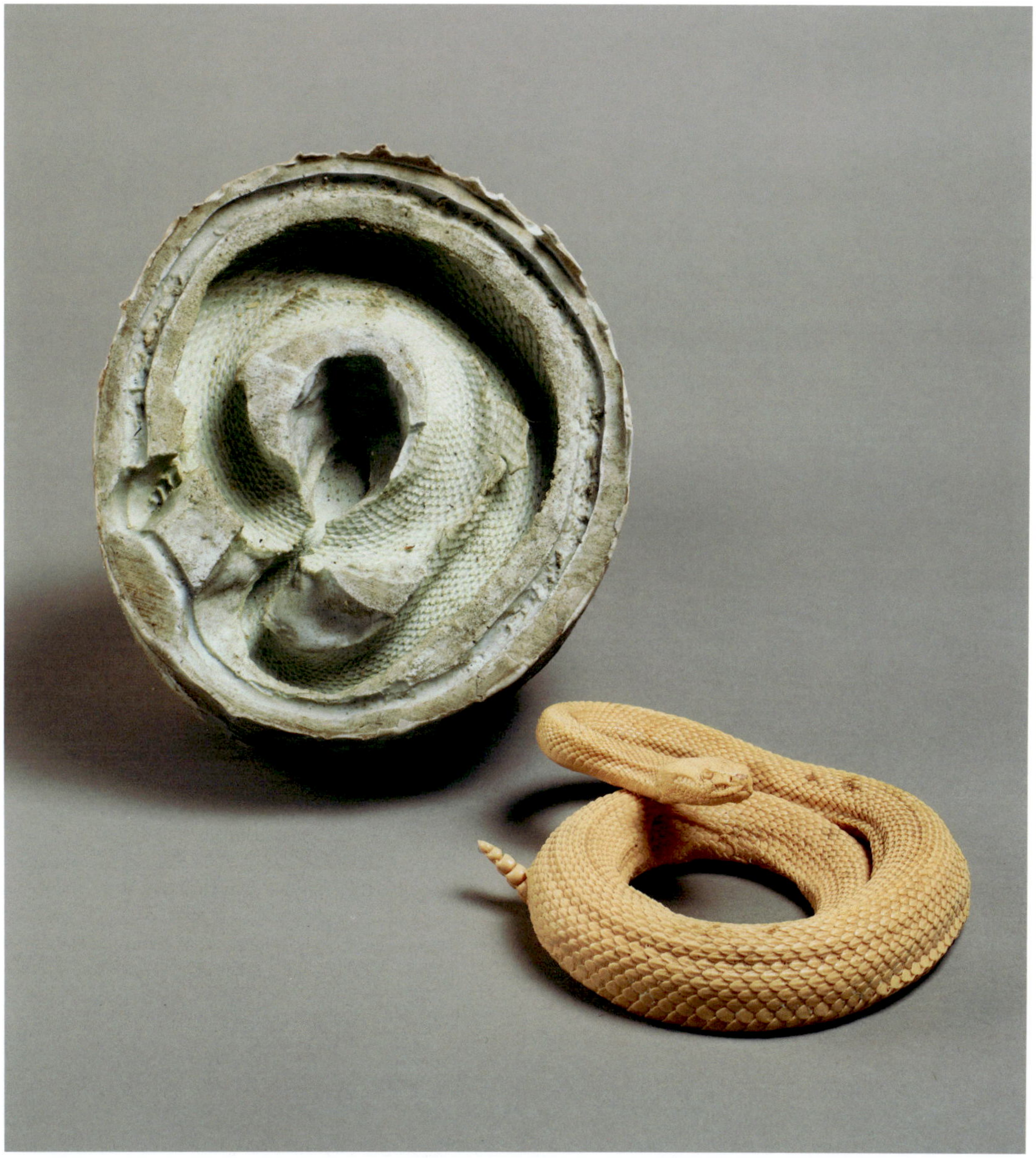

→ The skin of fish, amphibians, and reptiles does not preserve like that of mammals and birds, so these animals are not usually prepared in the same way. Rather than using real skin, a cast is made from a silicone mold of a real rattlesnake, to be painted later (2023).

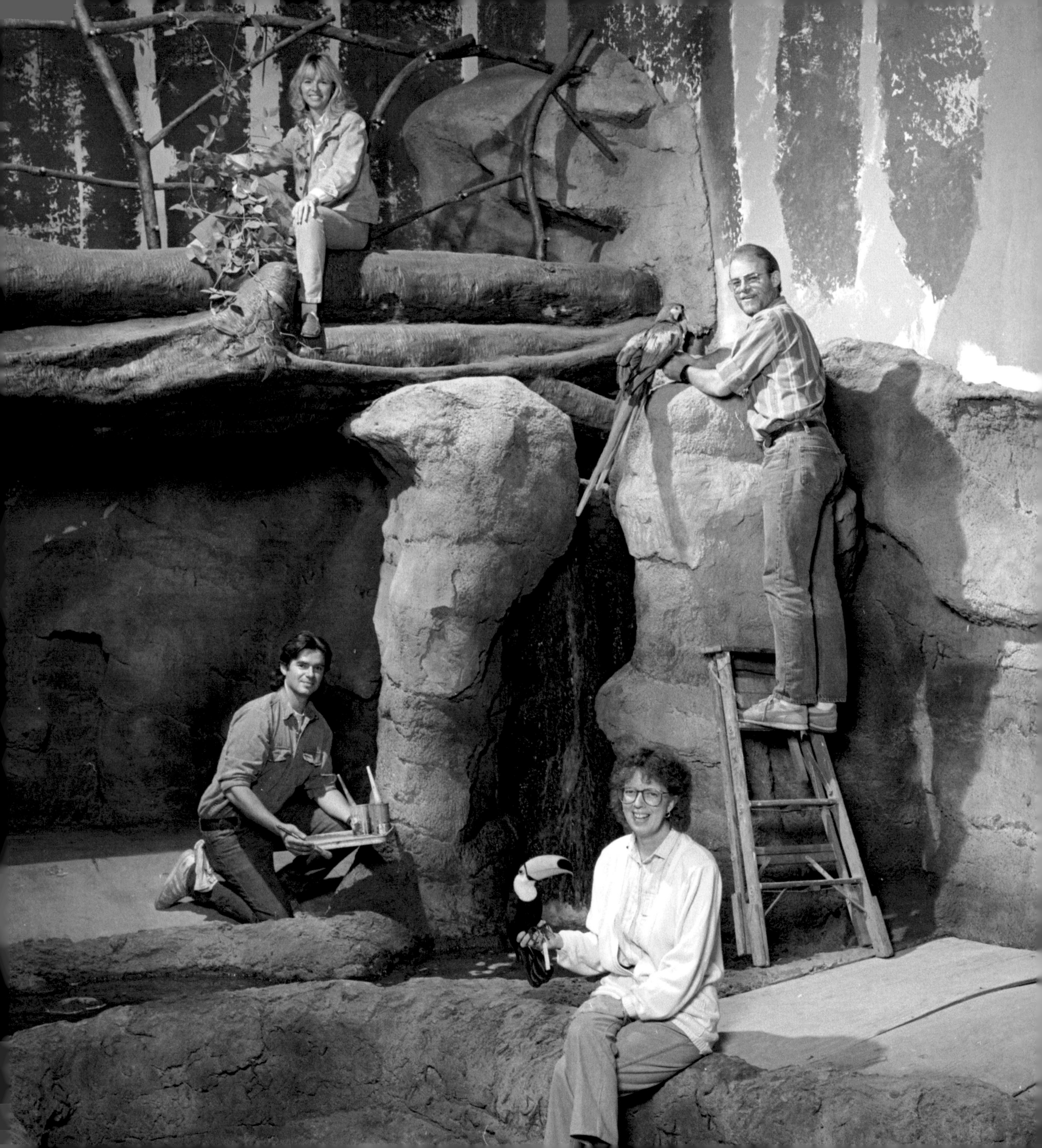

Assembly

Matt Davis

It takes the background artists, foreground artists, and taxidermists all working together as a team to create a captivating and realistic diorama. The background must blend into the foreground seamlessly, and all the animals need to be posed and positioned as if they were actually interacting with each other in the same space. During the assembly phase, all the different components of the diorama, from taxidermy to modeled plants and rocks, are installed and carefully positioned to create an engaging scene.

Large taxidermy, trees, and boulders are the first pieces to get moved into the diorama bay and bolted into place. Throughout most of the diorama halls' history, taxidermy was performed in a separate building across the street from the main museum. Finished mounts were trucked across the street and carried into the museum, sometimes right up the front steps.

Once large pieces are positioned, foreground artists start at the background painting and work toward the glass of the diorama, adding in plants, rocks, and small animals as they go. Desert and winter dioramas require an extra step to finish; workers must carefully sift new sand or (fake) snow onto the ground to cover up their footprints as they step out of the diorama and close up the glass.

← Each diorama requires a team of skilled artisans and scientists to assemble. *Clockwise from top left:* Caryl Castleberry, Tim Bovard, Betty Ann Schreiber, and Robert Reid pose while constructing the immersive *Rain Forest* diorama in the Ralph W. Schreiber Hall of Birds (1990).

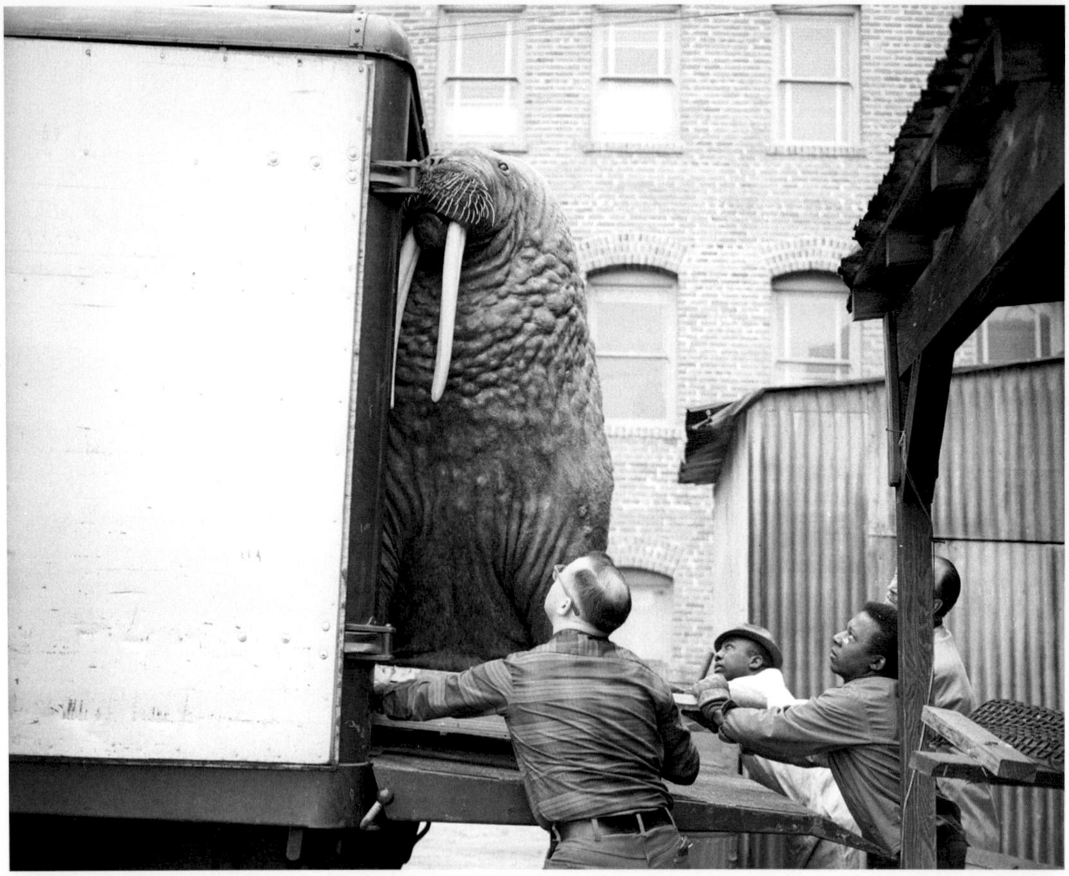

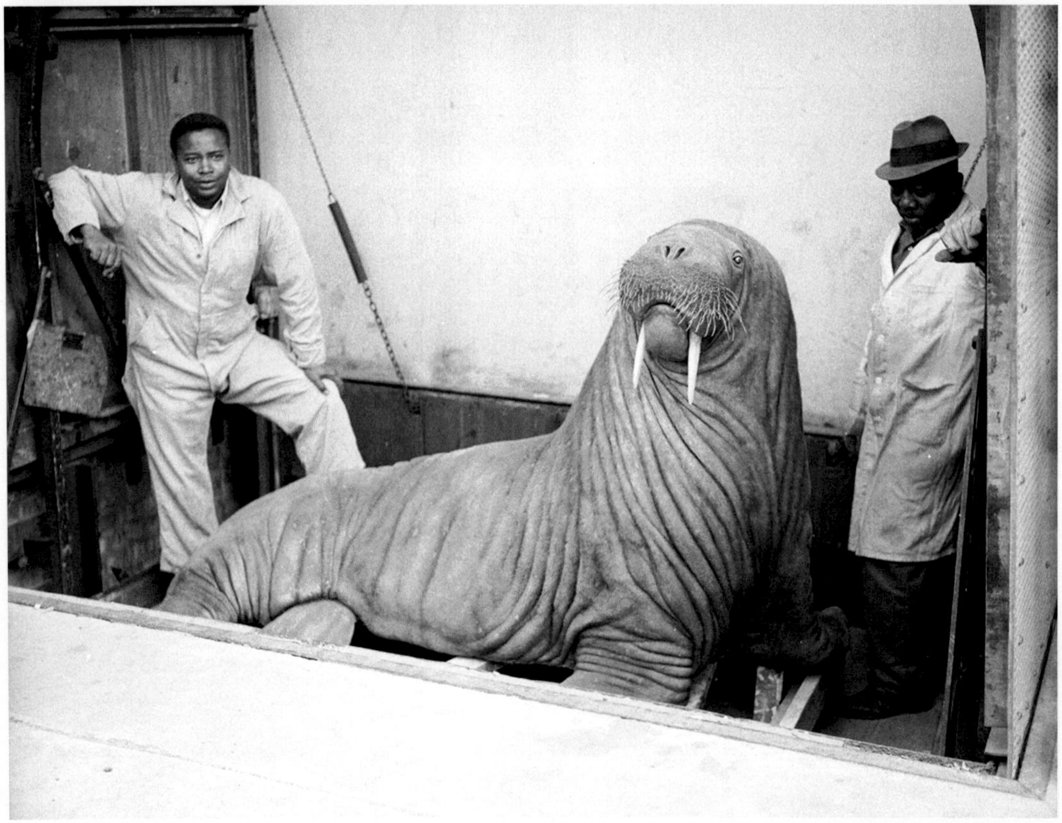

← Before the addition of a dedicated taxidermy and habitats studio on the fourth floor of the museum in the 1970s, the sometimes odorous work of taxidermy was performed in a separate building, just north of the main museum on what is now the campus of the University of Southern California. Once taxidermied animals like these walruses were ready to go into their dioramas, they had to be carefully loaded into a truck and driven across Exposition Boulevard before they could be carried into the museum (1967).

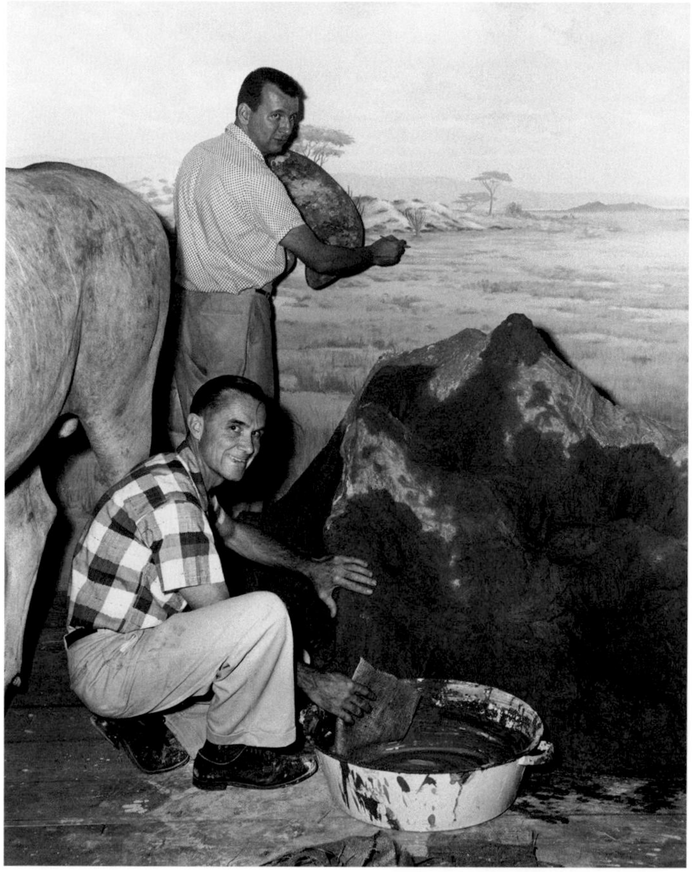

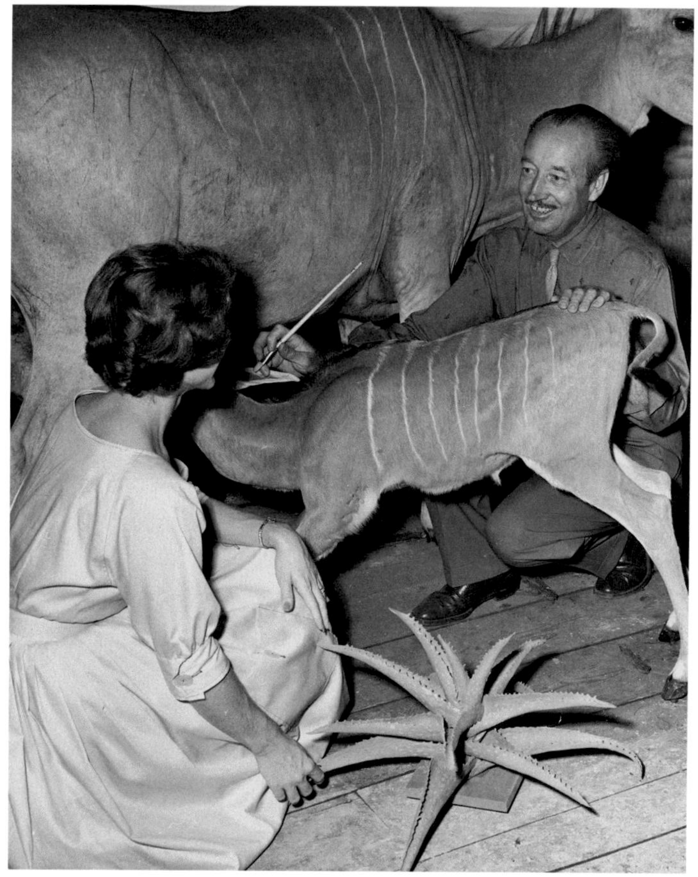

↑ Candid shots capture artists working to assemble version 2 of the *Eland* group (1960).

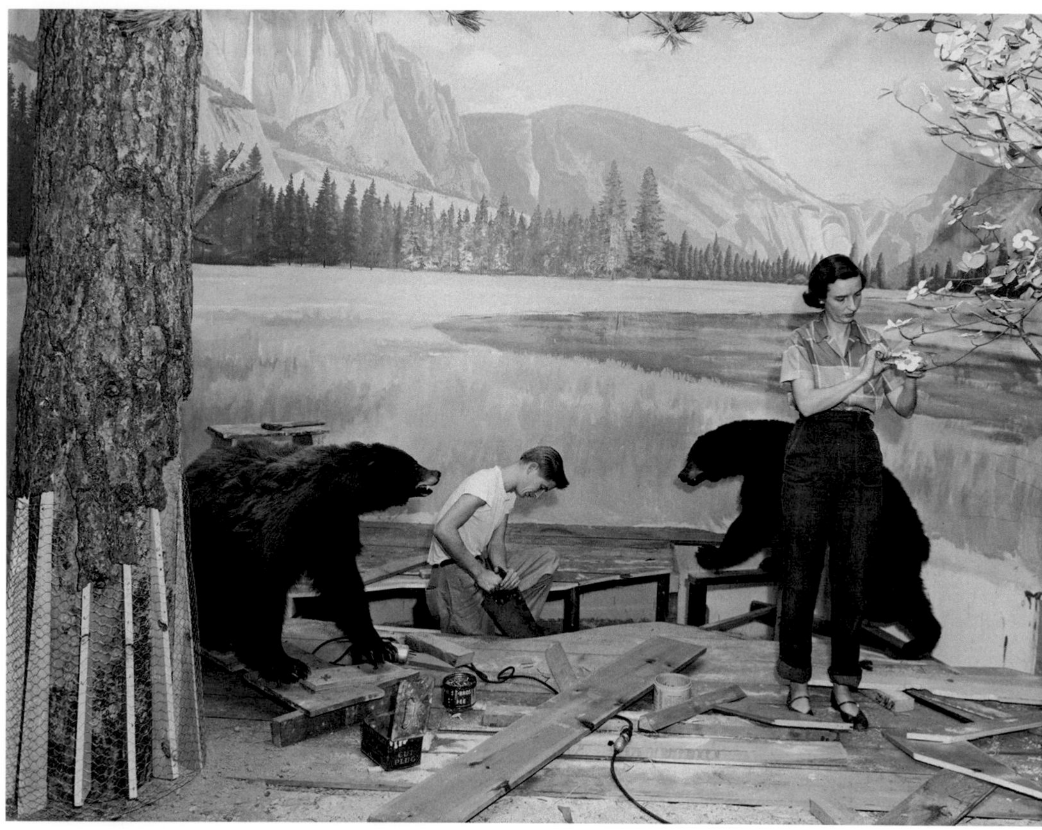

→ Version 1 of the *American Black Bear* group during assembly (c. 1950)

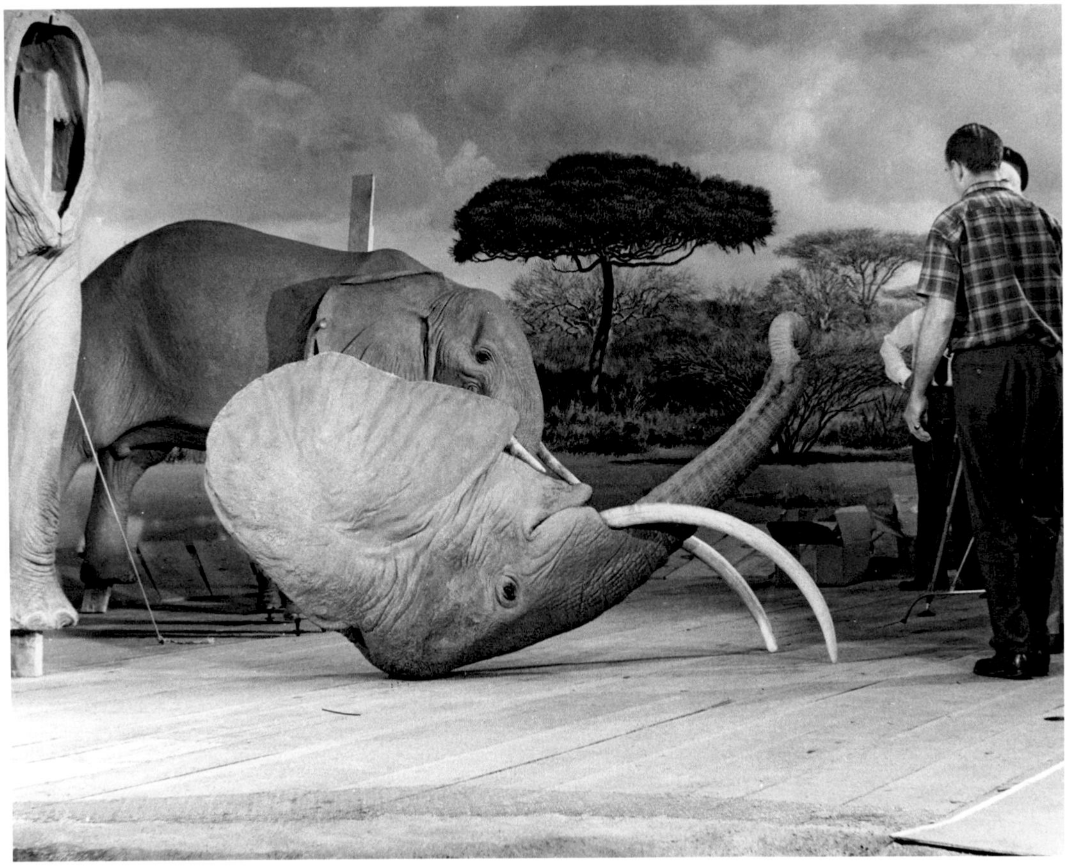

← The five African elephants collected for version 2 of the *Watering Hole* group were so large they had to be taxidermied off-site and carried into the museum. The bull elephant wouldn't fit through the door in one piece, so it was built in sections, lifted into the museum by a crane, and reassembled once inside the hall (*top left and bottom right*, 1966).

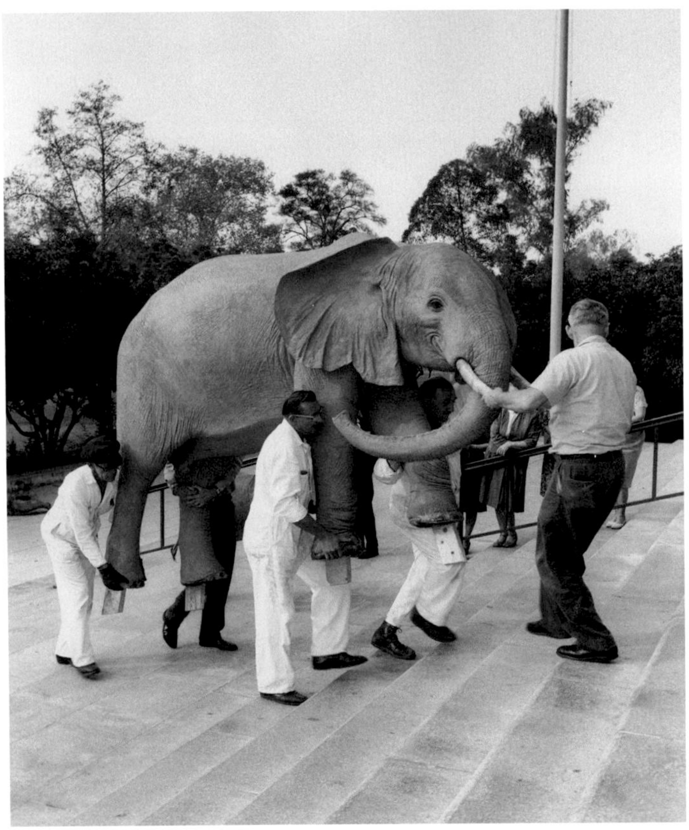

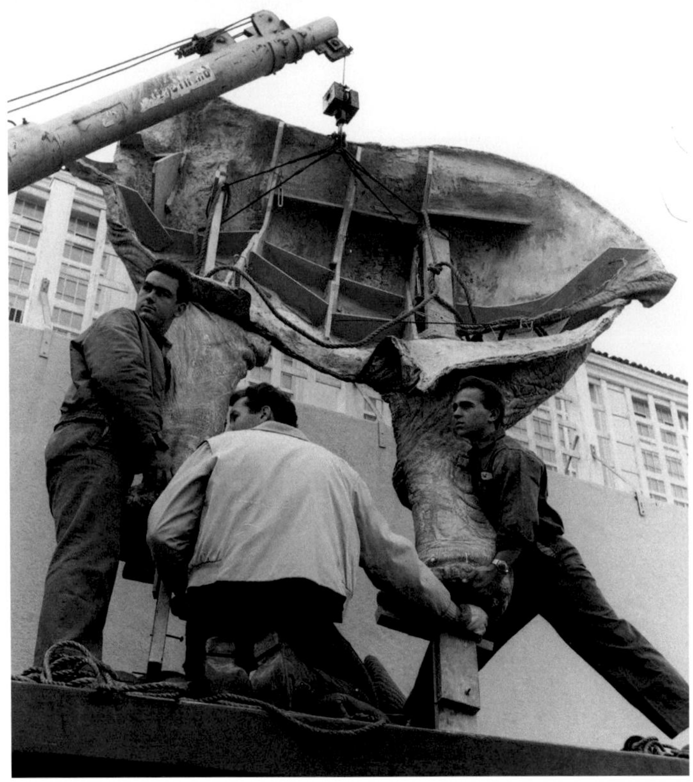

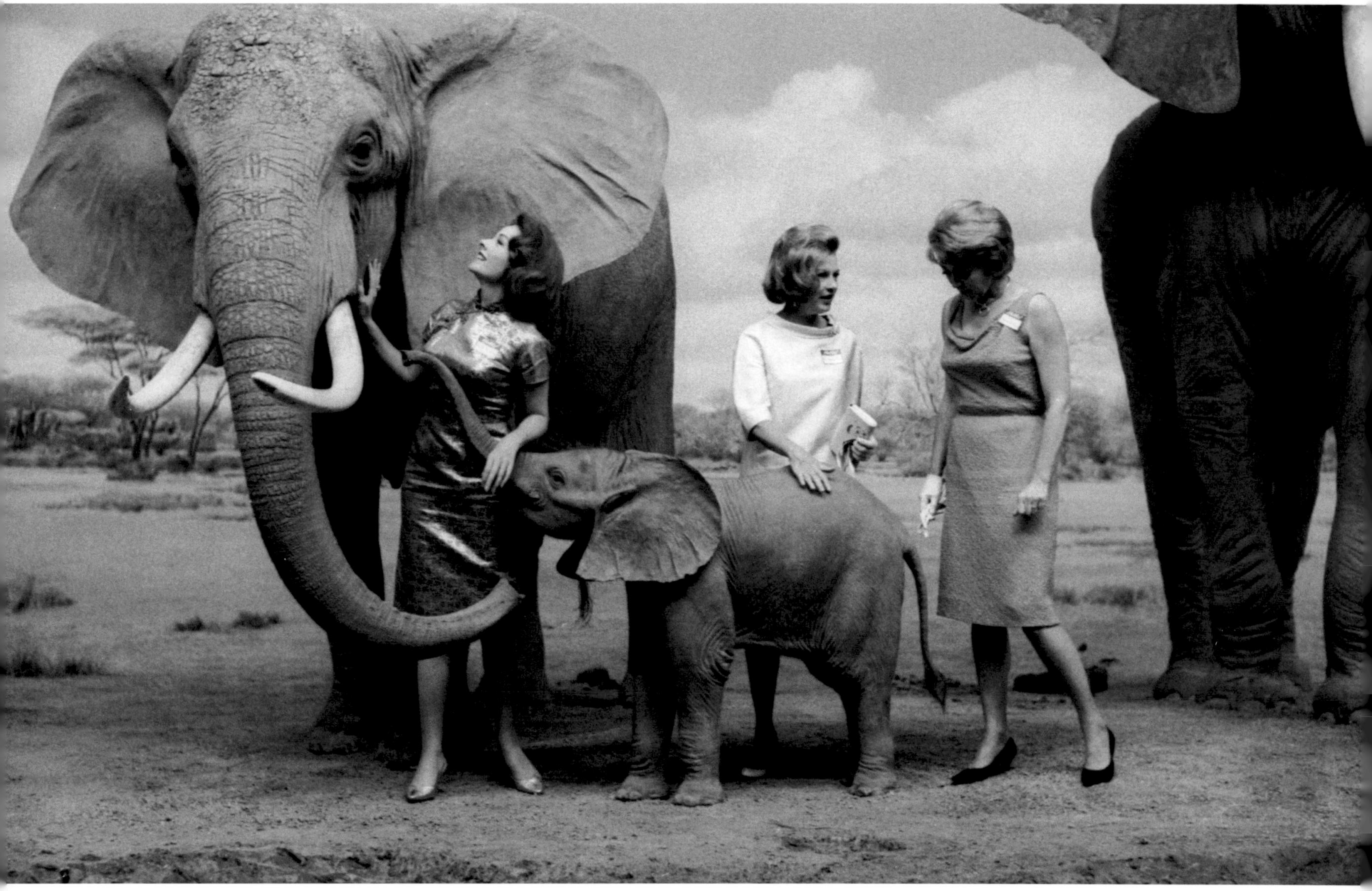

Upside down and sideways, in pieces and halves; headless and grotesque as they were at the time, it was all planned that way so that future generations may stand before the exhibit convinced that these great animals must have strolled in one day and suddenly froze in their tracks. It is a truly remarkable example of art and science combined in a single coordinated effort that makes this museum what it is today.[30]

—Taxidermist George Adams, 1966

↑ After assembly was complete, the museum opened its new *Watering Hole* diorama in 1966 with a gala affair complete with celebrities and dancers. Eager attendees were invited to walk into the diorama for pictures with the elephants.

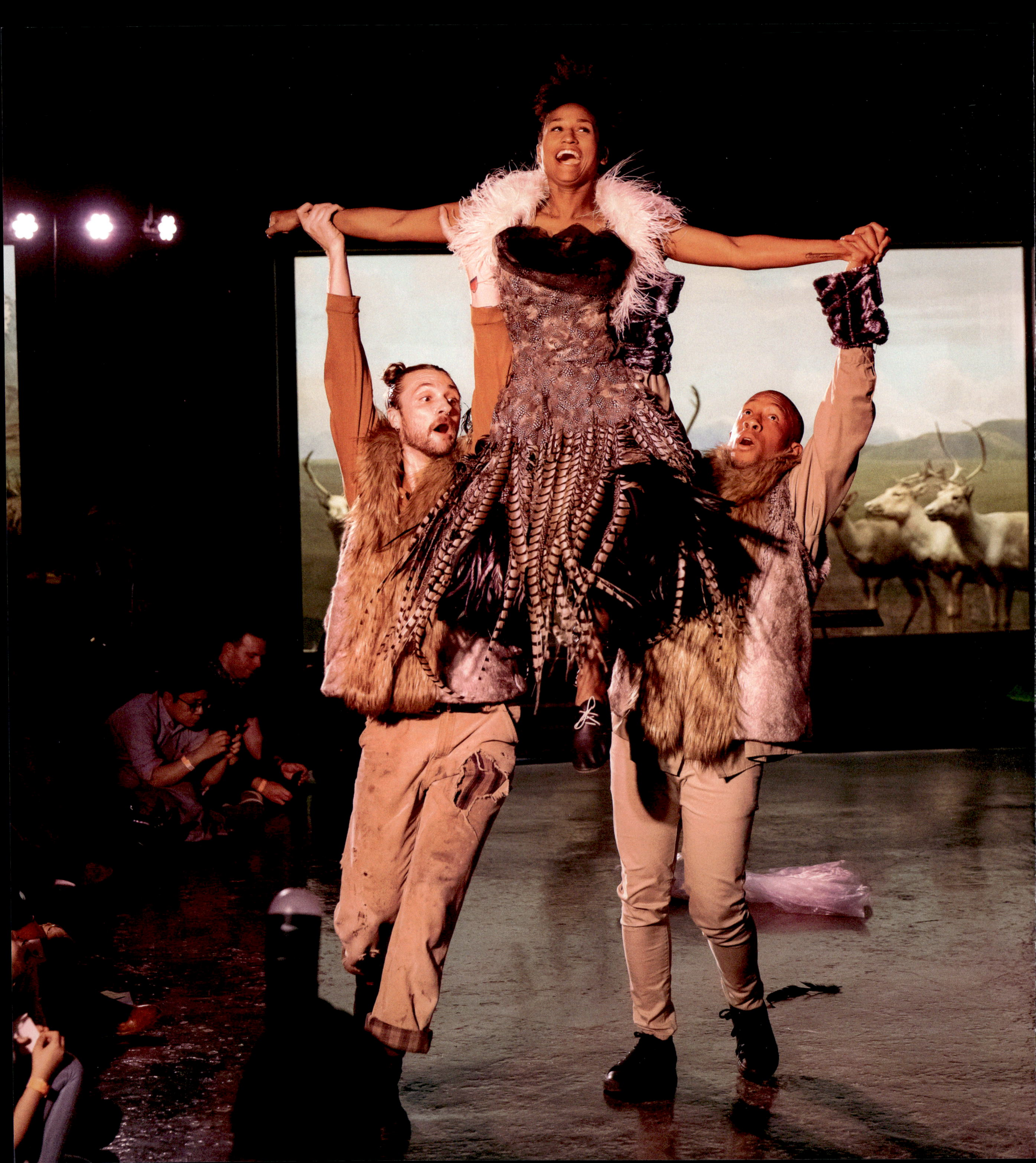

Conclusion

Matt Davis

← During Night of Ideas, the Heidi Duckler Dance Company performs a piece inspired by the animals surrounding them in the Level 1 North American Mammal Hall (2020).

In 1919, when he announced the creation of the new NHM diorama halls, Director Frank Daggett told the *Los Angeles Times* that he was not prepared to put any price on the groups then being assembled. "They are tremendously valuable now. In the future, they will be priceless."[31] Daggett was right; the dioramas (both their art and their specimens) are even more valuable today, over 100 years later. The museum continues to make new dioramas, but many aspects of the historical diorama halls would be impossible to replicate now. Few movies still use hand-painted backdrops, so it would be hard to find someone who could create a diorama background as quickly and beautifully as a skilled Hollywood painter like Duncan Spencer.

Some of the animals are now so rare that they would never be collected in the wild. Since the museum's *Black Rhinoceros* diorama opened in the 1950s, the black rhino population has plummeted over 94 percent from heavy poaching. Black rhinos are now recognized as critically endangered, which means that though the diorama may survive for another 50 years, the species inside it probably will not. Similarly, while the snow we see painted onto Mt. Kilimanjaro will last, given current climatic conditions, the glaciers of Africa's tallest mountain will likely completely melt away in the coming decades. This habitat

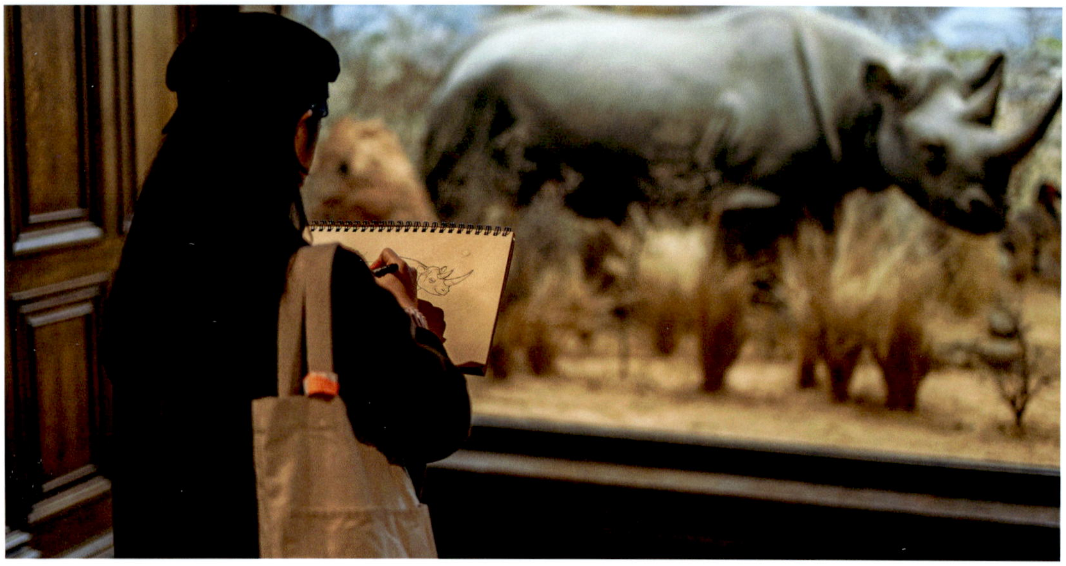

← A student on a field trip from Pasadena's ArtCenter College of Design draws inspiration from the *Black Rhinoceros* diorama and the rare species preserved inside (2023).

group is now dangerously close to fulfilling Taxidermist John Rowley's original vision of dioramas as permanent records of remarkable scenes that have since disappeared, a public reminder that such a majestic habitat once existed.

Just like the habitats and animals in them, dioramas themselves are deserving of preservation. They are incredible works of art that record a history of how museums have thought about biodiversity. Why value a diorama background painting by Hanson Puthuff any less than his small plein air paintings so sought after by collectors and art museums? Or why respect George Adams any less as a sculptor because he covered his lifelike art with real animal hides and called himself a taxidermist? It is hard to overstate what a revolutionary art form dioramas were. They were the vanguard of a movement that believed natural history museums should be for everyone, not just the specialist. The movement held that museums could better educate the public not through long boring labels but by inspiring visitors with captivating experiences. Dioramas were also the reflection of a major theoretical advancement in science, the recognition that everything in a habitat was connected, part of some greater whole, that the lowly blade of grass mattered as much as the massive elk stepping over it. Ecology is the philosophy embraced every time a child in grade school makes a shoebox diorama of an endangered species in its natural habitat.

Dioramas are also valuable because they preserve legacies that many would now rather forget. As scholar Donna Haraway recognized in 1985: "Dioramas are meaning-machines. Machines are time slices into the social organisms that made them."[32] For early diorama creators who feared many animals could not be saved from extinction, shooting a few condors and bison to preserve them in a museum for future generations made sense. But it doesn't make sense to NHM visitors anymore. The wealthy elite who hunted animals for the dioramas were celebrated for their civic-mindedness and generosity well into the 1960s, but most Angelenos would probably disapprove of their patronage today. A 2015 poll found that 86 percent of Americans opposed big game hunting.[33] The racist power structures in colonial Africa that made it possible for one man like Leslie Simson to legally shoot hundreds of animals for the museum become apparent when examining the country names on old diorama labels: Northern Rhodesia, Nyasaland, British East Africa. This colonial history is one that many museums are now struggling to address.

Even with this baggage, the dioramas' original message remains relevant: many animals and habitats are still threatened. No one shot the animals in the *Maned Wolf*, *Giant Armadillo*, and *Giant Anteater* dioramas; they were all donated by zoos. These species are at risk because they are losing their home, the Gran Chaco habitat in South America, which is an area larger than California. More than 25 percent of the remaining forest in the Gran Chaco has been destroyed since 2004, mostly to make room for cattle ranches and soybean farms to feed cattle. It's not trophy hunters or poachers that are threatening most animals, it's us and our insatiable desire for more land. Humans and our hungry livestock, like cows and pigs, now make up 96 percent of all mammals on earth (by weight). Diorama halls are the only places left

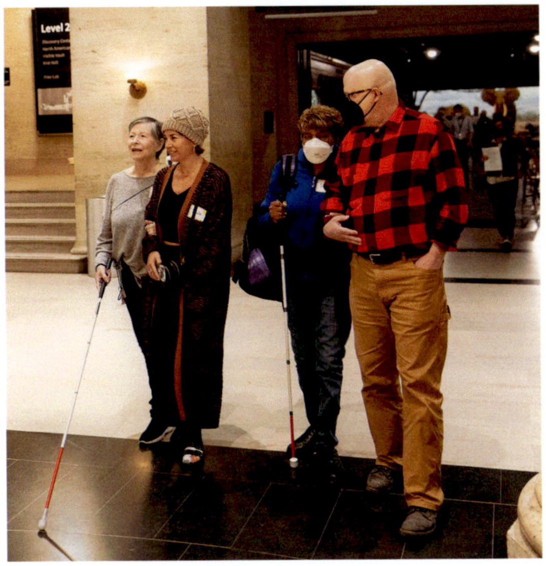

↑ NHM partners with organizations across Los Angeles to increase the accessibility of exhibitions. Visitors with visual impairments enjoy a guided tour of the dioramas led by sighted volunteers (2022).

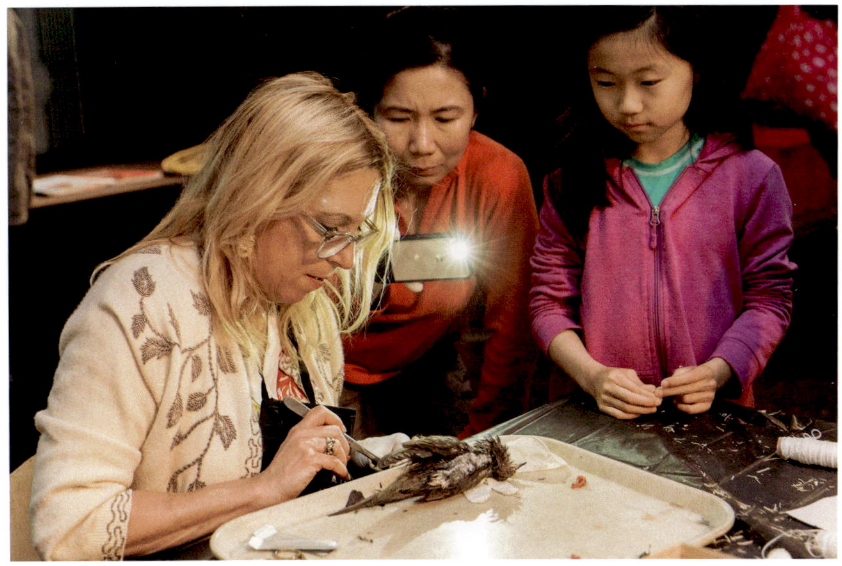

↑ NHM keeps the art of dioramas alive by holding demonstrations like this workshop, where staff and volunteers, including Gina Del Campo, show visitors the process of taxidermy up close (2018).

where wild animals still outnumber cows. To protect species today, we will have to make societal changes like the ones inspired by dioramas a century ago.

The artists and scientists who built NHM's first dioramas probably could not imagine what Los Angeles County would eventually become. Just as it has over the last century, NHM will continue to change and grow with the community it represents. As science and education develop, the museum will find new ways to communicate about biodiversity, but the dioramas will always have a special place in the hearts of Angelenos. As generations of visitors reflect on their first visits to these wonderful halls, we would be wise to consider the words Senior Curator of Habitats Melville Lincoln wrote to visitors back in 1941:

> So, to those great men of modern taxidermy who made possible the preservation of specimens of our disappearing wild life, particularly the larger mammals, we owe a debt of gratitude. It is our privilege and our obligation to insure that the museum will do its best to follow in their footsteps to present and preserve our wild life for the interest and education of the people.[34]

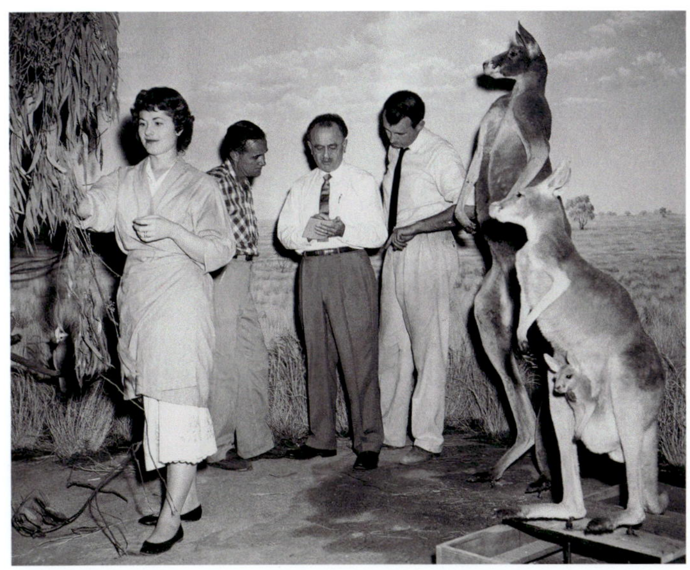

↓ Foreground Artist Peggy Clark adjusts leaves in the *Red Kangaroo* diorama while Senior Curator of Habitats Melville Lincoln (*center*) consults with Taxidermist Paul Hill (*left*) and Background/Foreground Artist Robert Clark (*right*) in 1958.

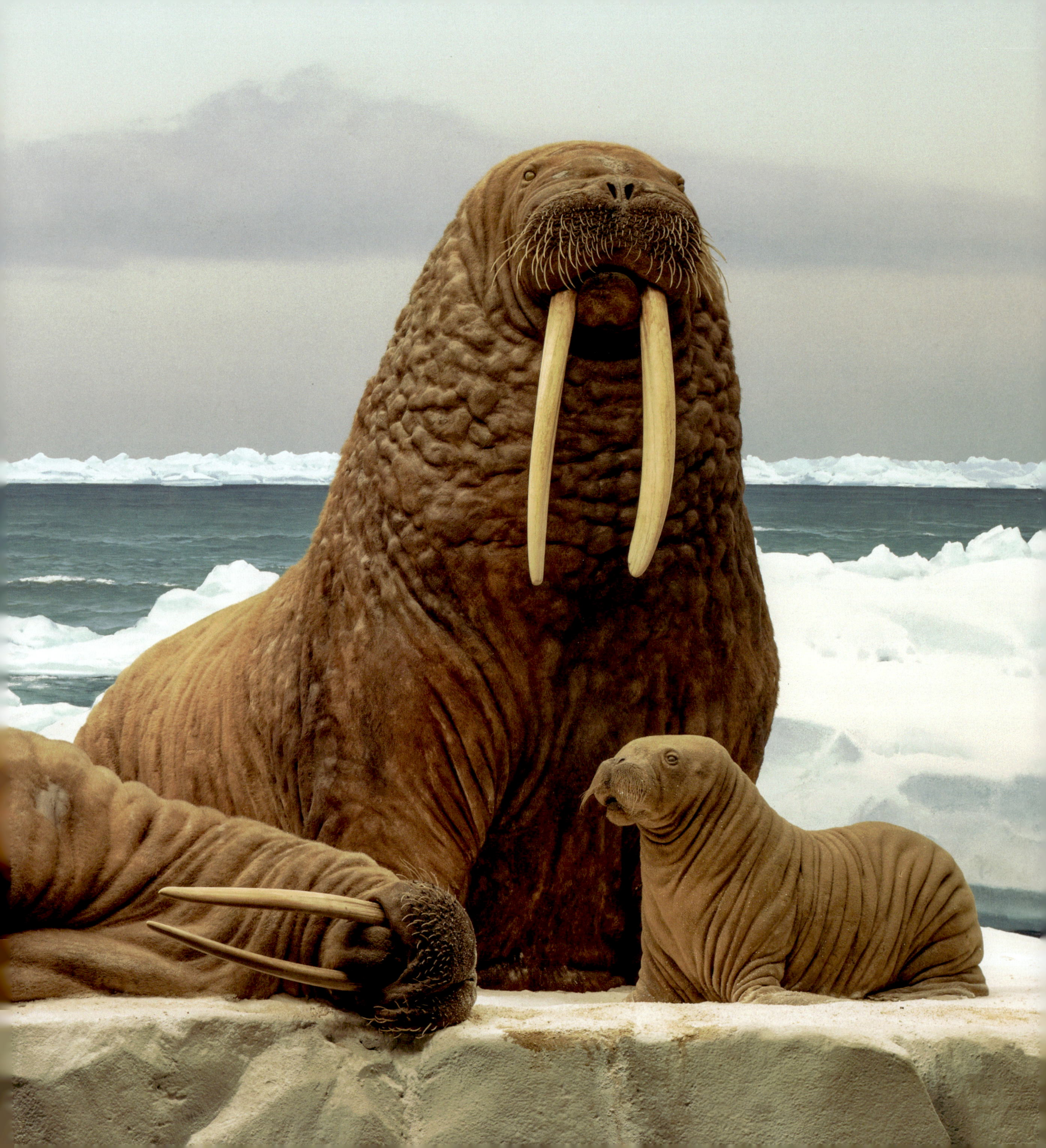

Appendix

The most extraordinary feature of the diorama halls at the Natural History Museum of Los Angeles County is how frequently the dioramas inside them change. If an elderly grandparent mentioned how much she loved the *Polar Bear* diorama, she could be referring to any of three different versions of that group, each with completely different foregrounds and backgrounds. These different versions were even in separate halls!

To clear up confusion, this appendix attempts to list every group that has been on display during the first 100 years of the diorama halls' history. This is no simple task. The dioramas were (and still are) continually updated, and not every change or opening was documented. The names of the halls themselves have changed over time. The McClellan Hall of North American Big Game eventually became the North American Mammal Hall (on Level 1) and then the even shorter M-B4, which was used for decades on old visitor maps that identified each gallery by a three-letter code. These codes can still be spotted in the diorama halls, hiding high up on the walls. A more informative approach may be to refer to the hall generically by its position, Level 1 East. Following are floor plans for each hall with their current names and some of the names they were previously known by.

← The *Walrus* group opened in 1967 in Bay 10 of the Level 2 Hall of North American Mammals. It is one of the few dioramas at NHM that has not changed drastically since it was first installed.

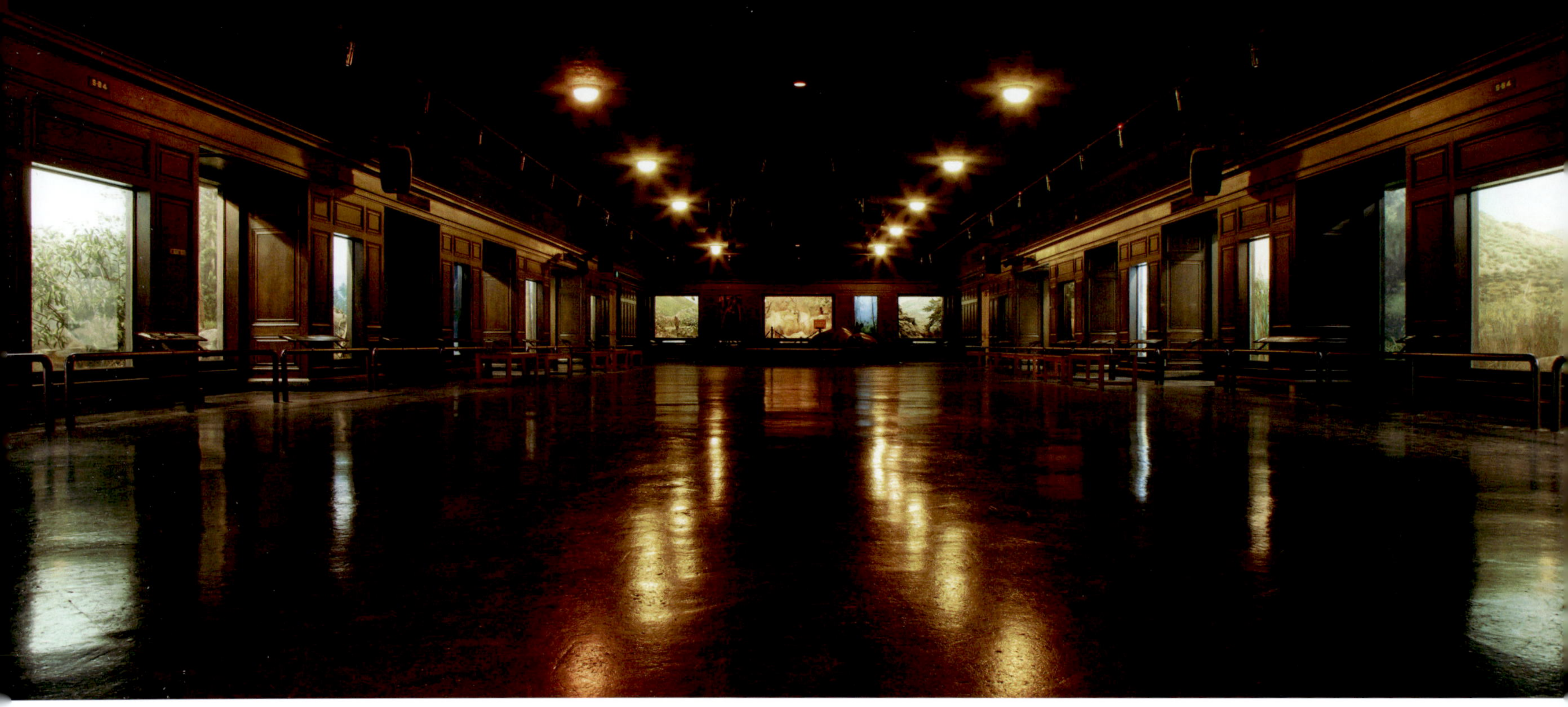

↑ The Level 2 North American Mammal Hall has 29 diorama bays, more groups than any other hall at the museum.

Thankfully, the curved diorama bays have not changed since the halls were first constructed. They are convenient reference points, numbered counterclockwise around each hall, starting with the bay on the right as one enters. After the floor plans comes a chart for each hall, showing which habitat groups were in each diorama bay and for how long. Although taxidermy is often added or altered in a group, updated versions signify major changes to a diorama, like a new background or foreground. Consider the elderly grandmother again. The dioramas she first visited as a child, though seemingly unchanged, are almost completely different than the dioramas her granddaughter would see on a school field trip today.

Floor Plans

Level 2

West

(wing that opened in 1930 as part of the Unit II Addition)

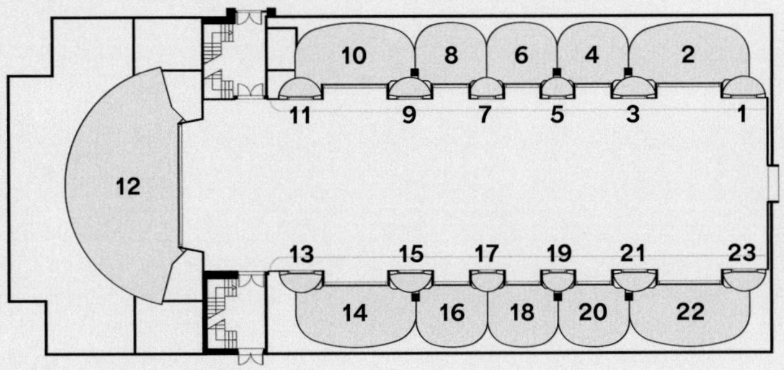

Level 2 West Hall

Previously known as
- Hall of Exotic Mammals
- Hall of African Mammals (Second Floor)
- S-B5

East

(wing that opened in 1925 as part of the Unit I Addition)

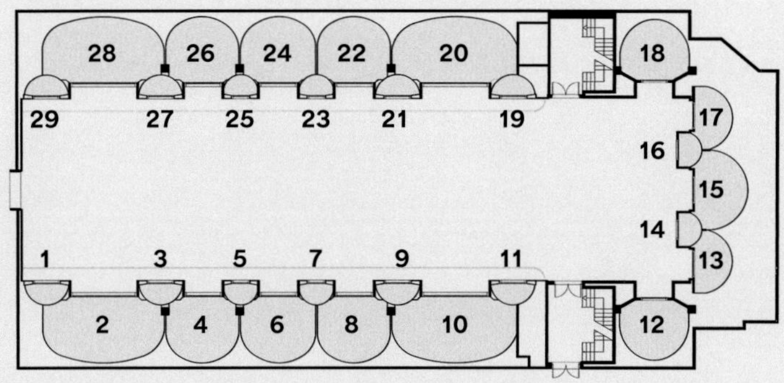

Level 2 North American Mammal Hall

Previously known as S-B4

Level 1

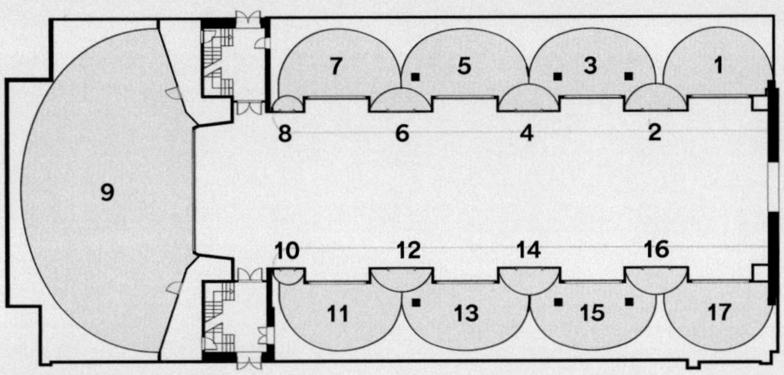

The Mr. and Mrs. Samuel Oschin and Family Hall of African Mammals

Previously known as
- The Simson Hall of African Mammals
- M-B5

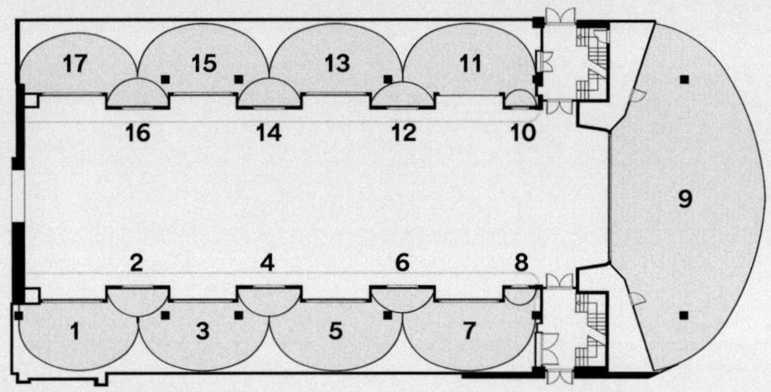

Level 1 North American Mammal Hall

Previously known as
- The McClellan Hall of North American Big Game
- M-B4

Level 1 North American Mammal Hall

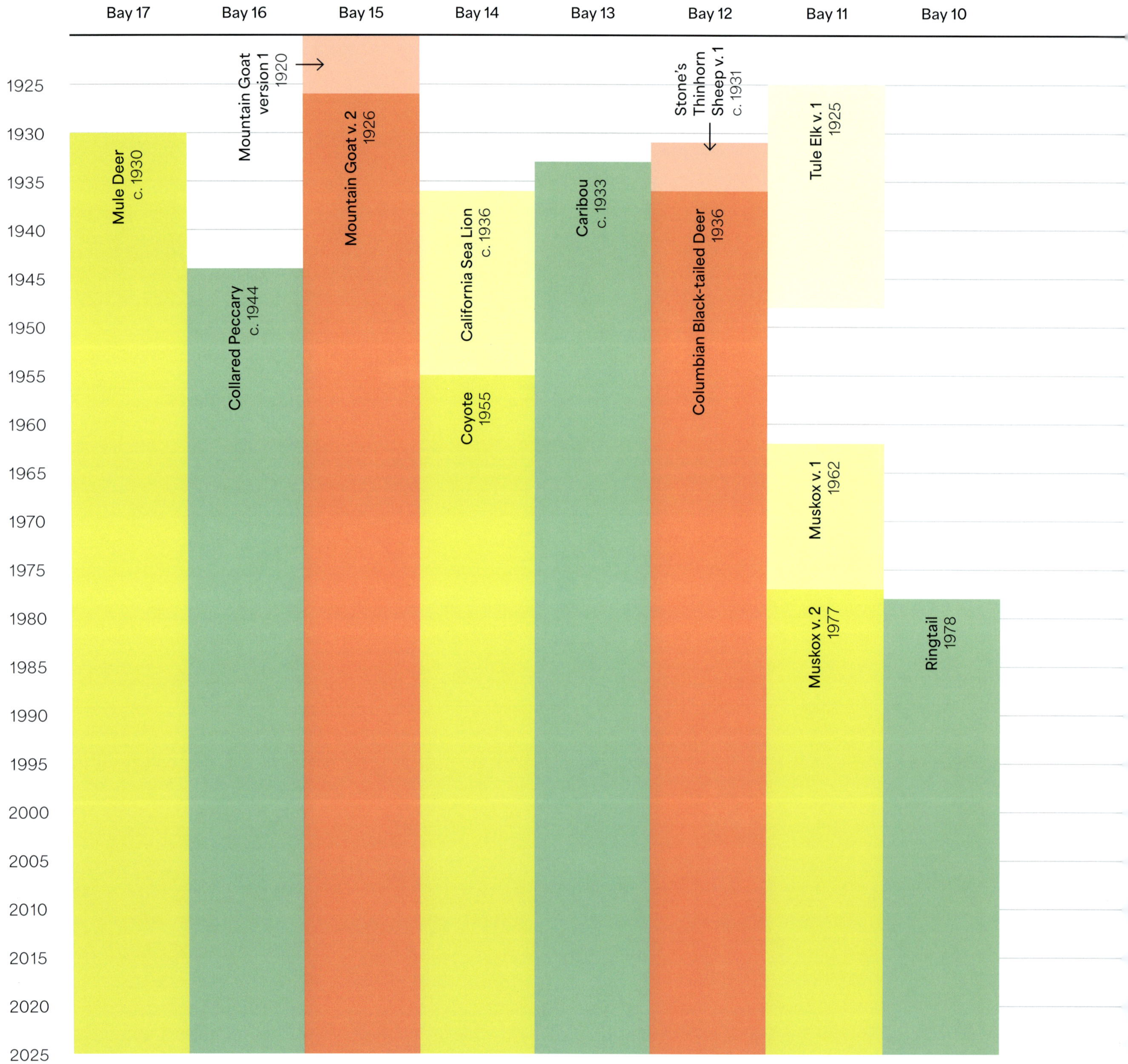

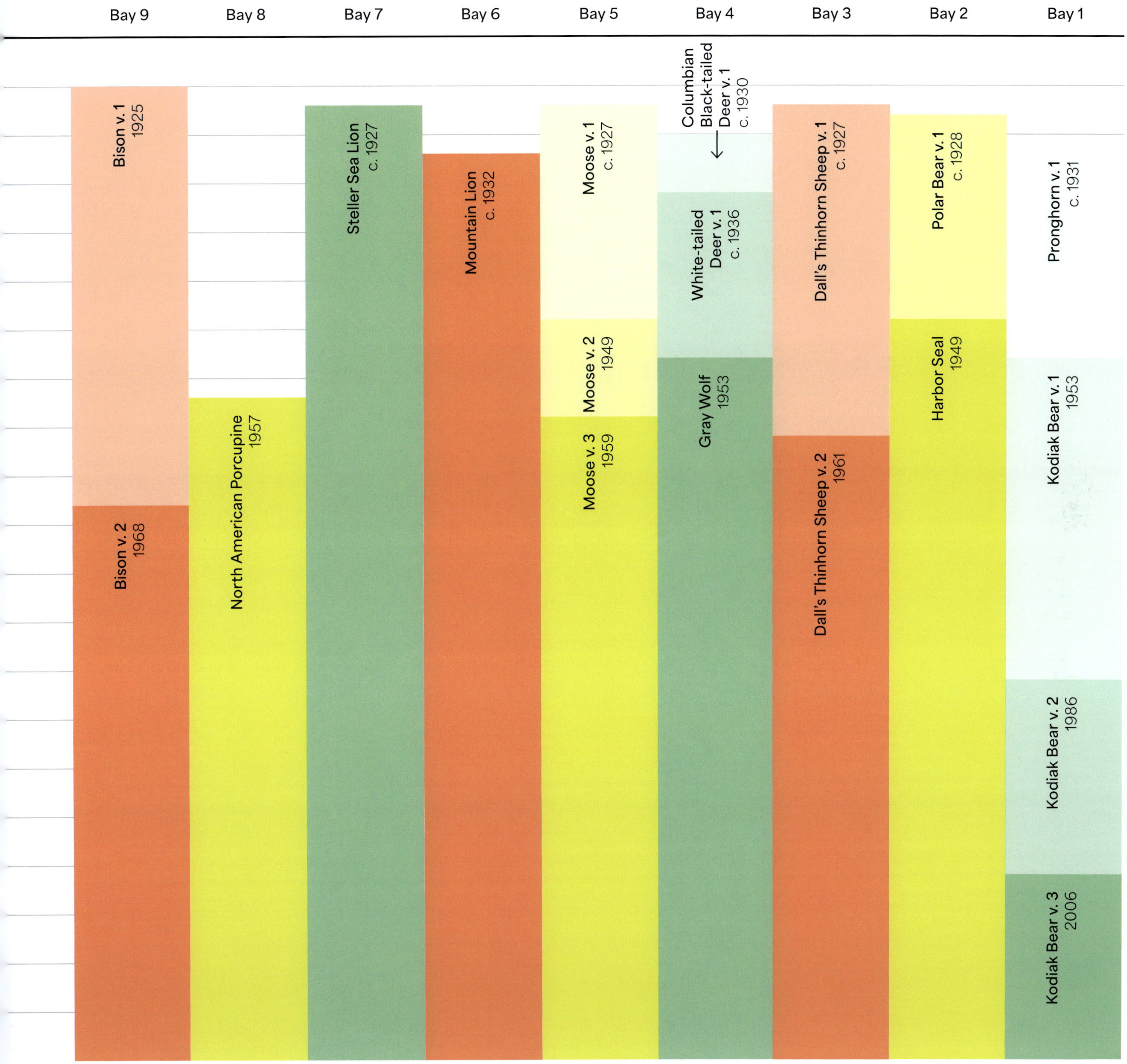
Bay 9
Bay 8
Bay 7
Bay 6
Bay 5
Bay 4
Bay 3
Bay 2
Bay 1
Bison v. 1 1925
Bison v. 2 1968
North American Porcupine 1957
Steller Sea Lion c. 1927
Mountain Lion c. 1932
Moose v. 1 c. 1927
Moose v. 2 1949
Moose v. 3 1959
Columbian Black-tailed Deer v. 1 c. 1930
White-tailed Deer v. 1 c. 1936
Gray Wolf 1953
Dall's Thinhorn Sheep v. 1 c. 1927
Dall's Thinhorn Sheep v. 2 1961
Polar Bear v. 1 c. 1928
Harbor Seal 1949
Pronghorn v. 1 c. 1931
Kodiak Bear v. 1 1953
Kodiak Bear v. 2 1986
Kodiak Bear v. 3 2006

Level 2 North American Mammal Hall

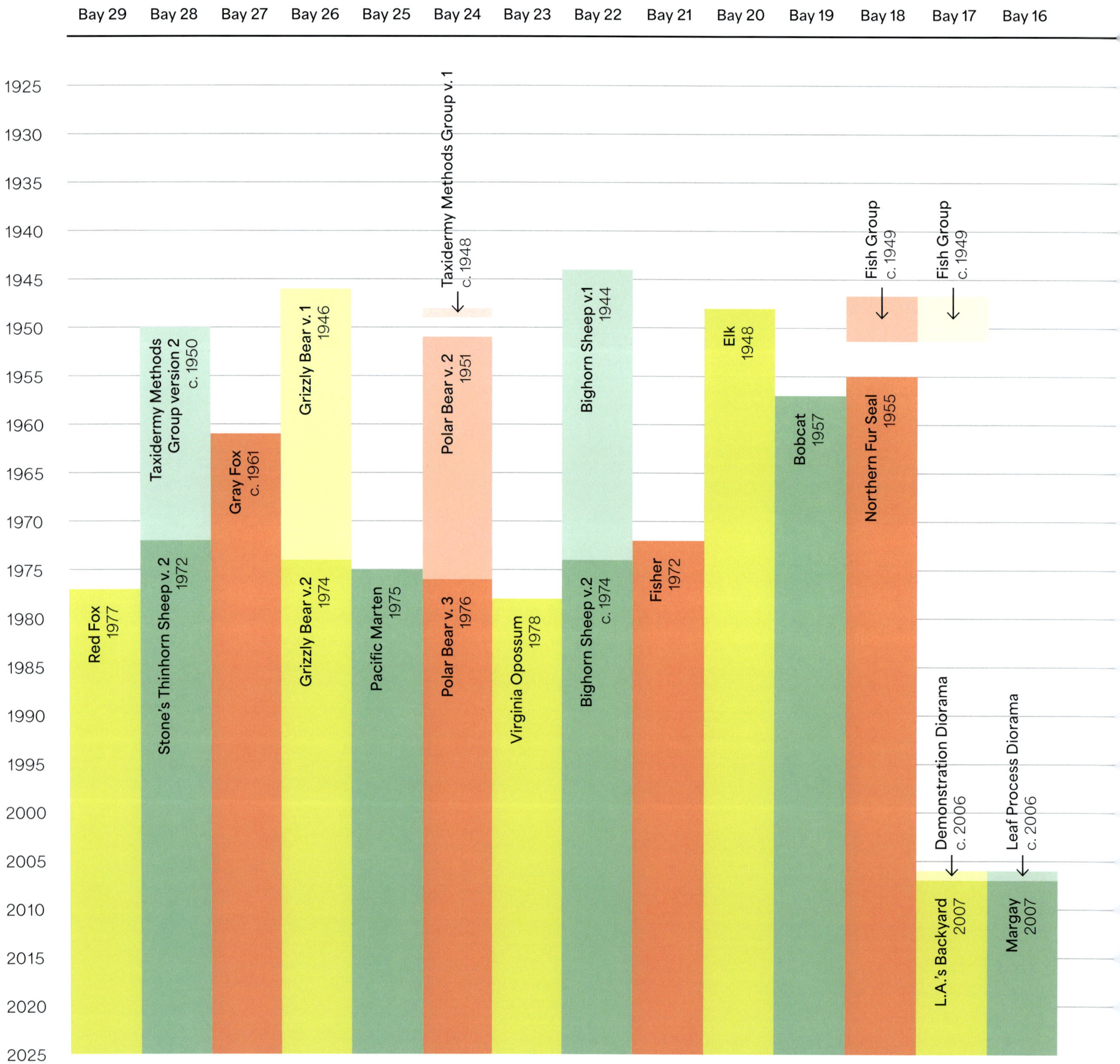

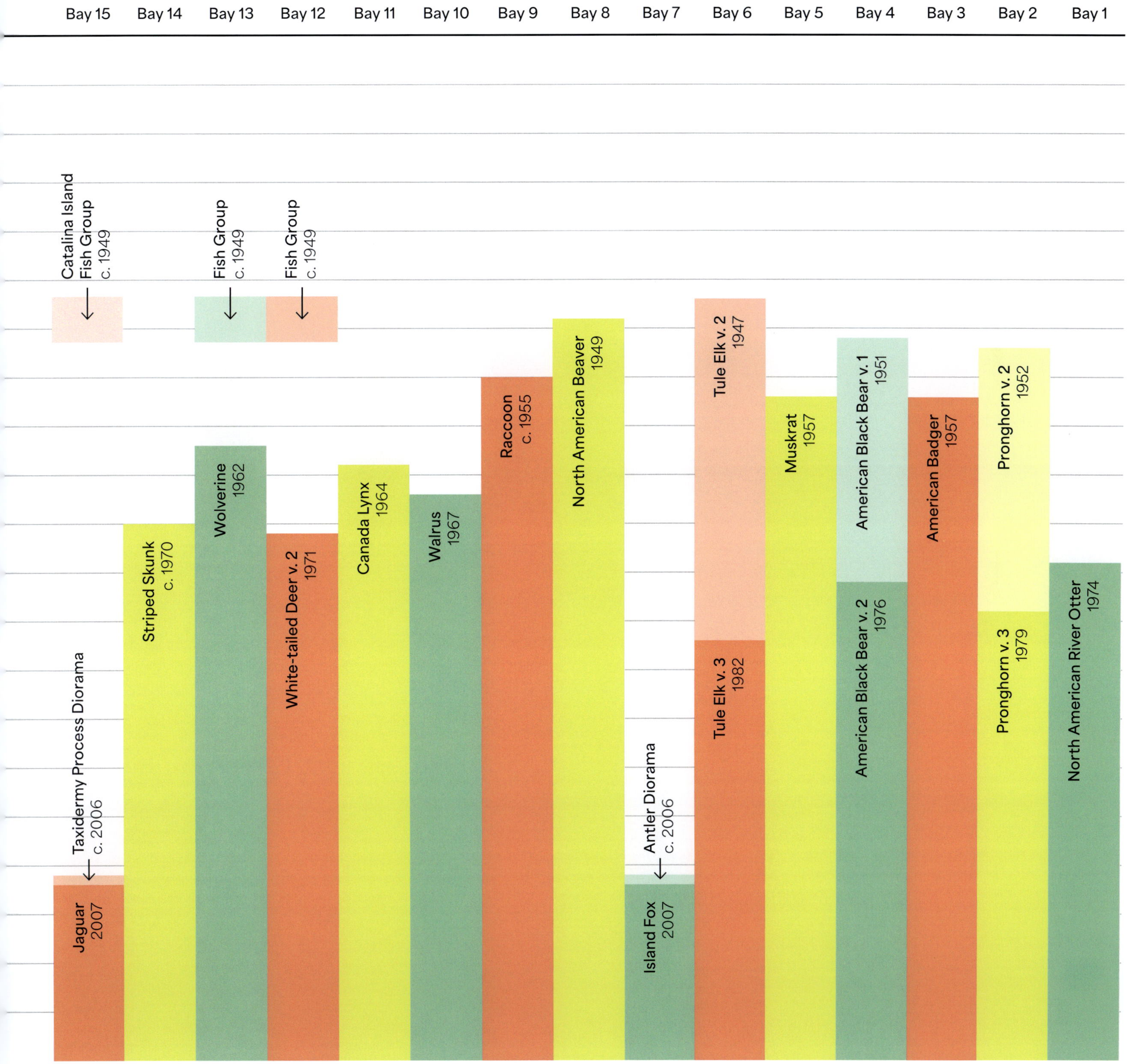

Bay 15
Bay 14
Bay 13
Bay 12
Bay 11
Bay 10
Bay 9
Bay 8
Bay 7
Bay 6
Bay 5
Bay 4
Bay 3
Bay 2
Bay 1
Catalina Island Fish Group c. 1949
Fish Group c. 1949
Fish Group c. 1949
Taxidermy Process Diorama c. 2006
Jaguar 2007
Striped Skunk c. 1970
Wolverine 1962
White-tailed Deer v. 2 1971
Canada Lynx 1964
Walrus 1967
Raccoon c. 1955
North American Beaver 1949
Antler Diorama c. 2006
Island Fox 2007
Tule Elk v. 2 1947
Tule Elk v. 3 1982
Muskrat 1957
American Black Bear v. 1 1951
American Black Bear v. 2 1976
American Badger 1957
Pronghorn v. 2 1952
Pronghorn v. 3 1979
North American River Otter 1974

The Mr. and Mrs. Samuel Oschin and Family Hall of African Mammals

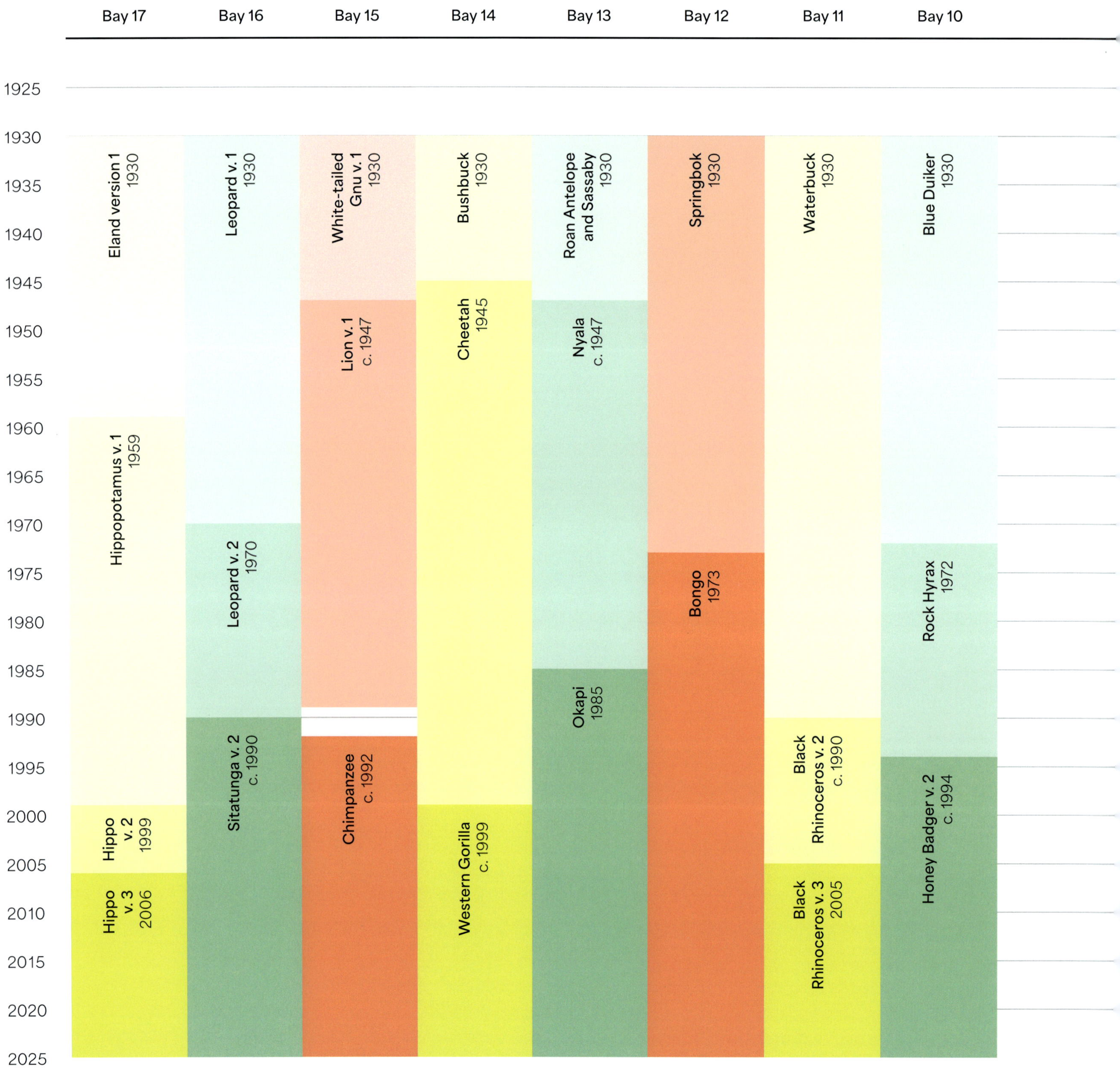

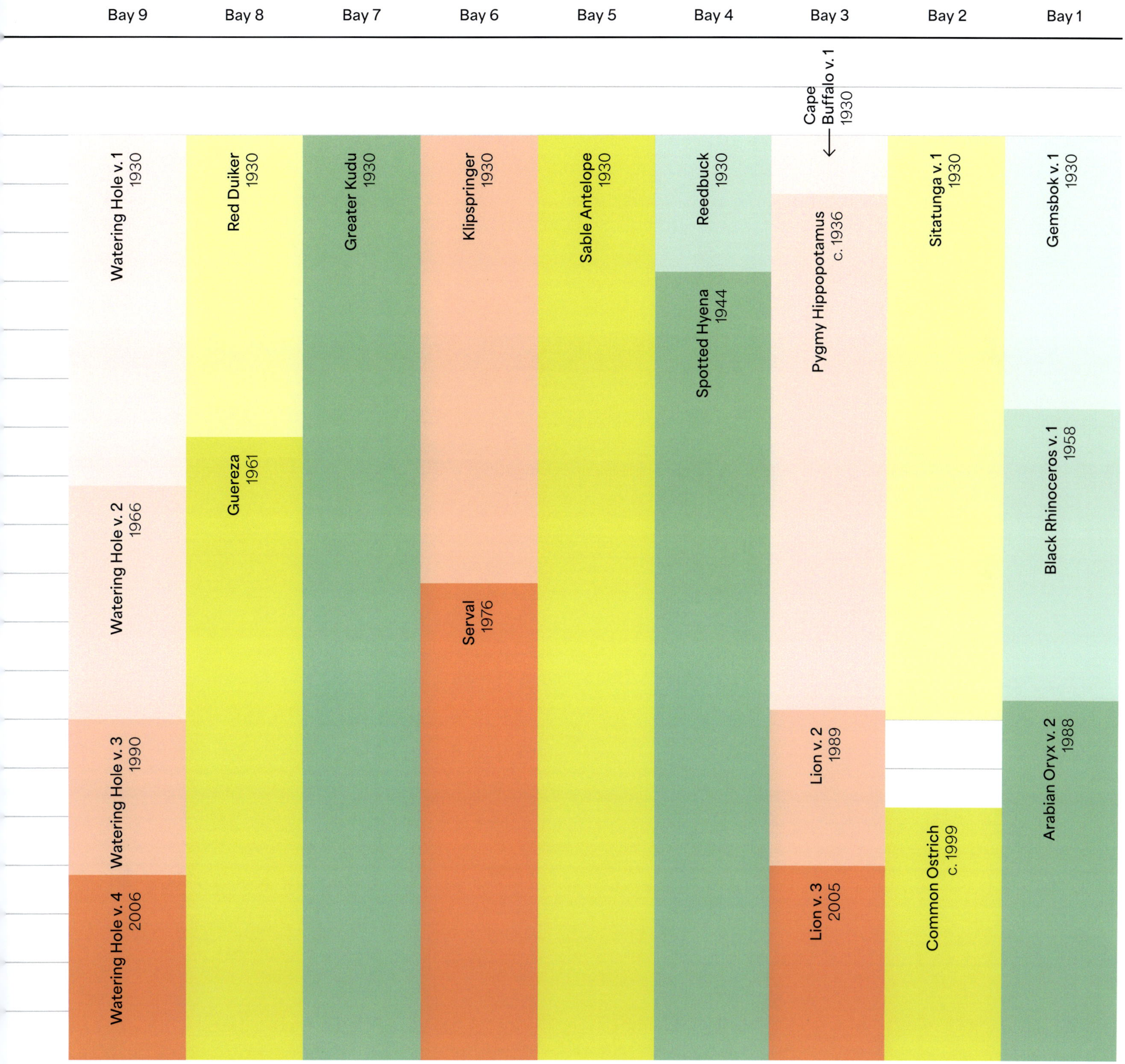

Bay 9
Bay 8
Bay 7
Bay 6
Bay 5
Bay 4
Bay 3
Bay 2
Bay 1
Watering Hole v. 1 1930
Watering Hole v. 2 1966
Watering Hole v. 3 1990
Watering Hole v. 4 2006
Red Duiker 1930
Guereza 1961
Greater Kudu 1930
Klipspringer 1930
Serval 1976
Sable Antelope 1930
Reedbuck 1930
Spotted Hyena 1944
Cape Buffalo v. 1 1930
Pygmy Hippopotamus c. 1936
Lion v. 2 1989
Lion v. 3 2005
Sitatunga v. 1 1930
Common Ostrich c. 1999
Gemsbok v. 1 1930
Black Rhinoceros v. 1 1958
Arabian Oryx v. 2 1988

Level 2 West Hall

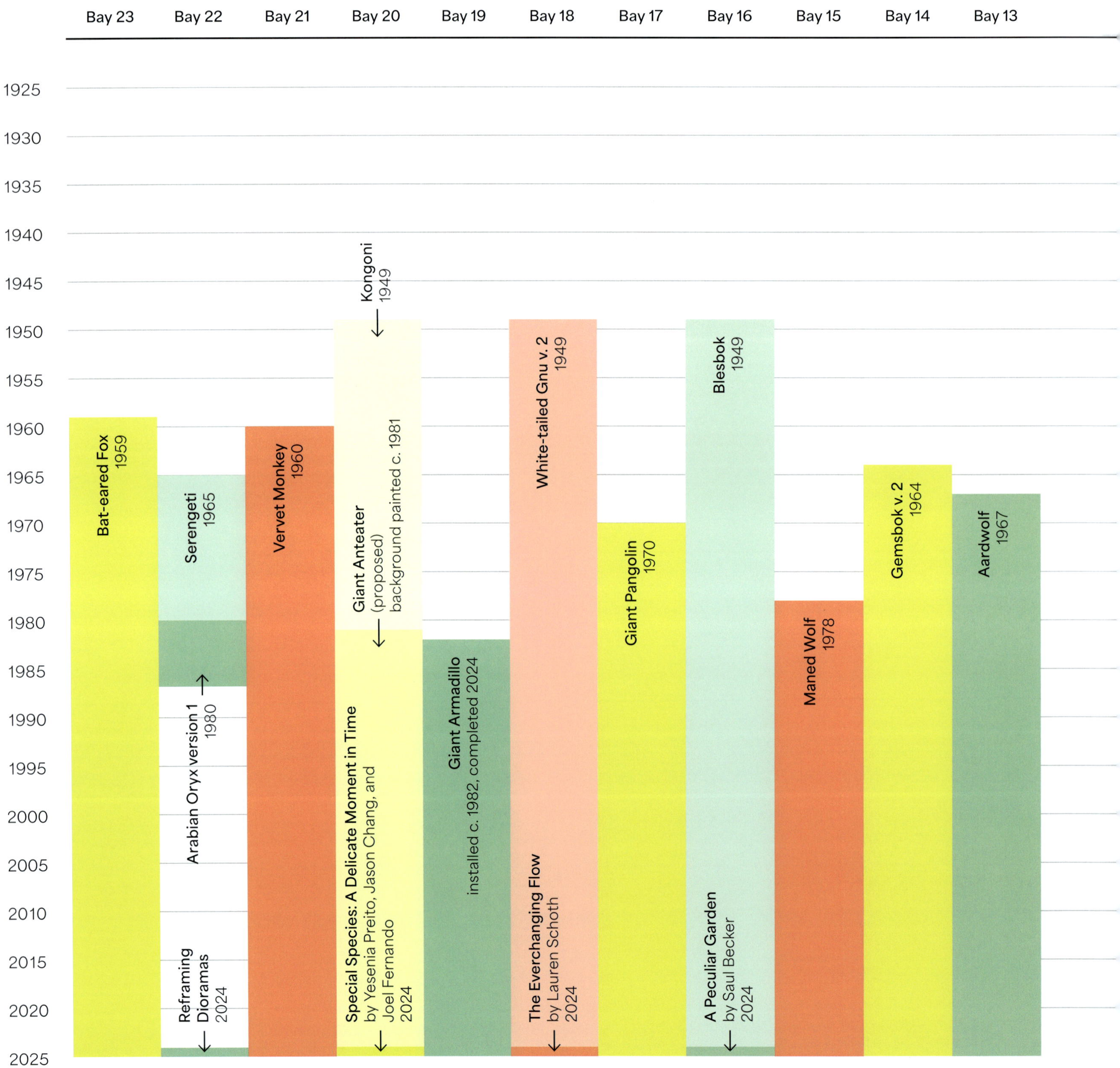

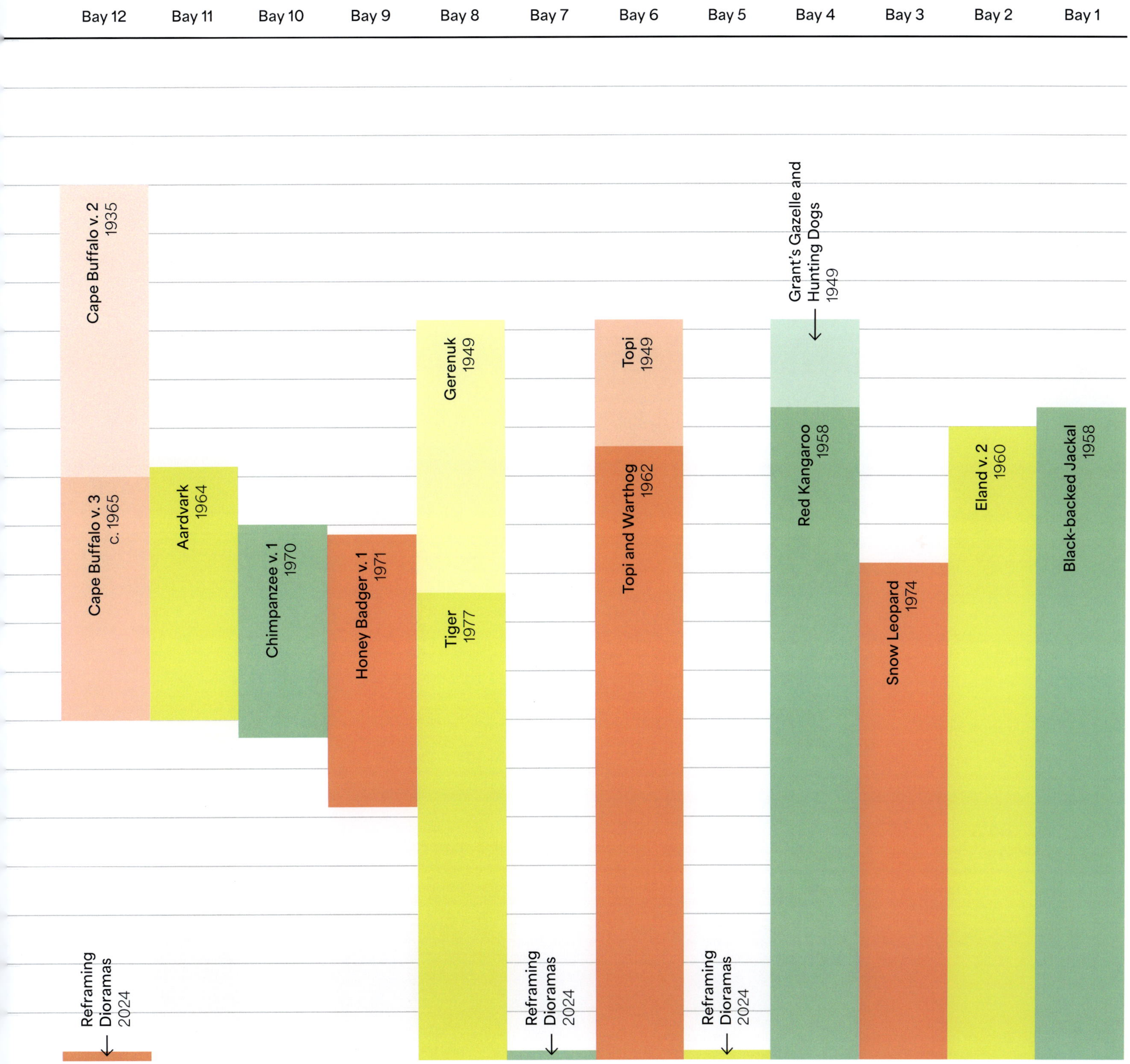

Bay 12
Bay 11
Bay 10
Bay 9
Bay 8
Bay 7
Bay 6
Bay 5
Bay 4
Bay 3
Bay 2
Bay 1
Cape Buffalo v. 2 1935
Cape Buffalo v. 3 c. 1965
Aardvark 1964
Chimpanzee v. 1 1970
Honey Badger v. 1 1971
Gerenuk 1949
Tiger 1977
Topi 1949
Topi and Warthog 1962
Grant's Gazelle and Hunting Dogs 1949
Red Kangaroo 1958
Snow Leopard 1974
Eland v. 2 1960
Black-backed Jackal 1958
Reframing Dioramas 2024
Reframing Dioramas 2024
Reframing Dioramas 2024

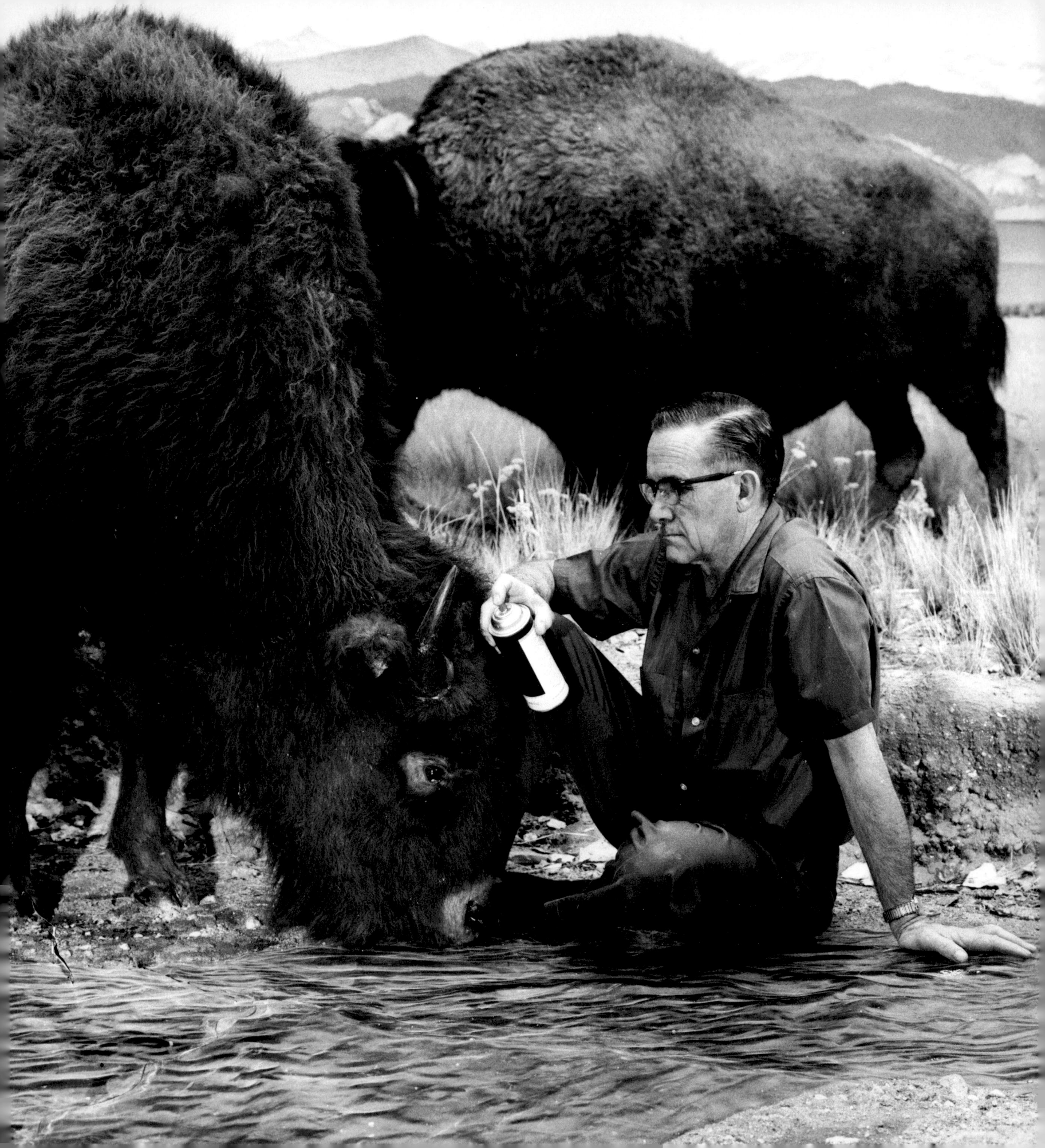

Acknowledgments

Research and publication for this book were generously supported by two grants from Getty through its PST ART: *Art & Science Collide* initiative and three Getty Marrow Undergraduate Internships. The authors would like to thank the many people involved in the creation of this book from conception through archiving and research to writing and editing. Marlyn Musicant, Liz Andres, Agnes Novie, and Jessica Sanchez provided project management. Yolanda Bustos, Cyrene Cruz, and Jenn Berger preserved and organized archives. Rose Keiko Higa and Christina Marsh assisted with research. Donna Pungprechawat, David Pullido, and Kristie Poon helped with image formatting and design. Natalja Kent, Liam Mooney, Chris Candelaria, K. T. Hajeian, and Betty Uyeda assisted with photography. Tania Collas and Marina Gibbons conserved specimens, and Su Oh, Chris Weisbart, and Andrea Mode Andrews provided executive leadership.

← Taxidermist Paul Hill touches up a bison (1970).

Notes

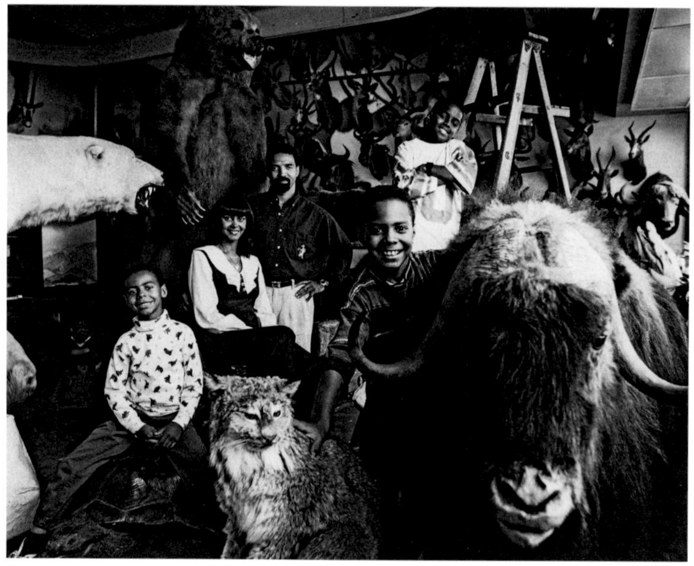

↑ The Crenshaw family poses in the museum's Taxidermy Studio for a 1997 membership promotion. *From left:* Ryker, Vivienne Swanigan-Crenshaw, Michael, Michael Jr., and Marchean.

1. Museum Director William A. Bryan to Dr. W. H. Osgood of Field Museum, January 1933, NHM.ARC.001, Directors Files, box 19, folder 5, item 01, Frank S. Daggett and William A. Bryan Collection, NHM Archives and Special Collections, Natural History Museum of Los Angeles County (NHM), Los Angeles, California.

2. Director Frank S. Daggett to Board of Supervisors regarding specimens brought to museum from Alaskan expedition, October 21, 1919, main filing cabinet, box 8, folder 234, item 56, Tim Bovard Taxidermy Collection, NHM Archives and Special Collections.

3. Melville Lincoln, "History of the Habitat Group," *Los Angeles County Museum Quarterly* 1, no. 3 (1941): 22.

4. Homer Dill and William Bryan, "Report on an Expedition to Laysan Island in 1911: Under the Joint Auspice of the United States Department of Agriculture and the University of Iowa," *Bulletin of the Biological Survey of the U.S. Department of Agriculture* 42 (1912): 1-30.

5. John Rowley, "The Proposed African Mammal Hall for Los Angeles Museum," *Museum Graphic* 1, no. 2 (1926): 48.

6. Homer Dill, "Building an Educational Museum as a Function of the University," *Tenth Annual Meeting of the American Association of Museums*, ed. Paul M. Rea (San Francisco: Waverly Press, 1915), 84.

7. William T. Hornaday, *Taxidermy and Zoological Collecting: A Complete Handbook for the Amateur Taxidermist, Collector, Osteologist, Museum-Builder, Sportsman, and Traveller* (New York: Charles Scribner's Sons, 1894), 222.

8. *Ferris Bueller's Day Off*, directed by John Hughes (Paramount Pictures, 1986), 103 minutes.

9. Claude Lévi-Strauss, *Le totémisme aujourd'hui* (Paris: Presses Universitaires de France, 1962), 112. Translation from the French by the authors.

10. *Mad Men*, season 5, episode 12, "Commissions and Fees," created by Matthew Weiner (Santa Monica, CA: Lionsgate Television, 2012), DVD.

11. Lévi-Strauss, *Le totémisme aujourd'hui*.

12. News release dictated to Supervisor Dodge by Museum Director Frank Daggett, June 1919, NHM.ARC.001, Directors Files, box 3, folder 33, items 01-05, Frank S. Daggett and William A. Bryan Collection, NHM Archives and Special Collections.

13. Samuel Hubbard, "Where Science Joins Hands with Art: The Modern Taxidermy as Revealed in San Francisco's Great New Museum," *Scientific American* 116, no. 6 (February 1917): 154-57.

14. Acting Museum Director to California Fish and Game Commission, November 13, 1920, NHM.ARC.001, Directors Files, box 4, folder 17, items 11-12, Frank S. Daggett and William A. Bryan Collection, NHM Archives and Special Collections.

15. H. E. Anthony of American Museum of Natural History to Museum Director William A. Bryan, June 18, 1931, NHM. ARC.002, Directors Files, box 15, folder 24, item 02, Frank S. Daggett and William A. Bryan Collection, NHM Archives and Special Collections.

16. Ralph Morrill to William H. Schuchardt, Report on Dioramas, October 28, 1949, NHM. ARC.005, Directors Files, box 31, folder 10, items 01-11, James A. Breasted Jr. Collection, NHM Archives and Special Collections.

17. Morrill to Schuchardt, Report on Dioramas.

18. "Fete to Mark Showing of Rhinoceros Display," *Los Angeles Times*, March 27, 1958.

19. Herbert Friedmann, "The Artistry of the Habitat Group,"

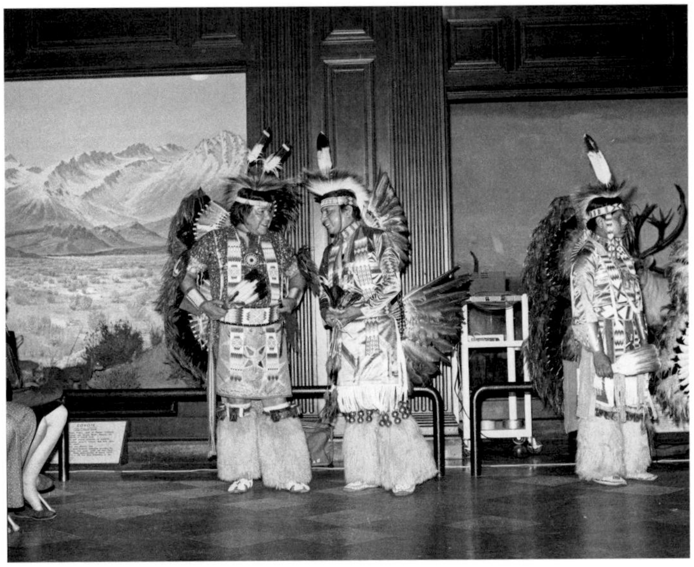

↑ Dancers prepare for a performance at a powwow hosted in the Level 1 North American Mammal Hall (1980).

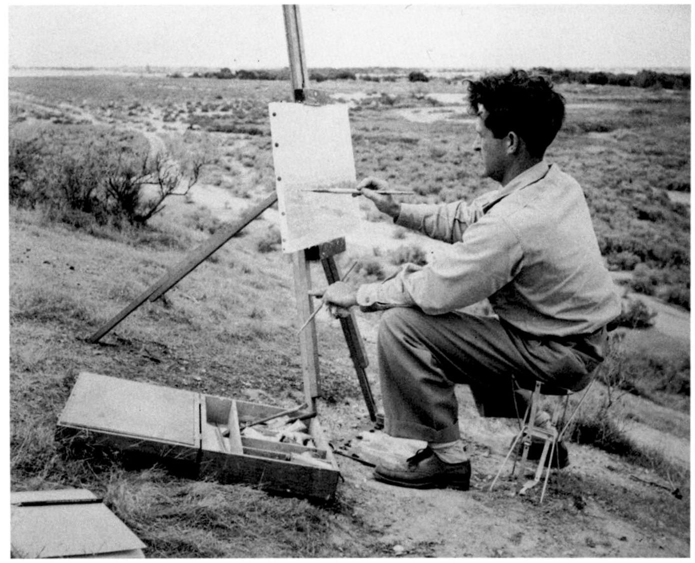

↑ Staff Artist J. Robert "Bob" Sewell paints a location study for version 2 of the *Tule Elk* group. A gifted painter and model maker, he left the museum to work on the Grand Canyon diorama at Disneyland. He would join Disney full-time as one of the earliest Imagineers and is honored with multiple gag tombstones at the Haunted Mansion attraction (1940s).

Los Angeles County Museum Quarterly 3, no. 1 (1964): 5-9.

20. James D. Olson and Donald R. Patten, "Museum-Zoo Relations and the Arabian Oryx," *Terra* 19, no. 1 (1980): 19-24.

21. Rowley, "Proposed African Mammal Hall," 44.

22. Joseph Grinnell, "The Uses and Methods of a Research Museum," *Popular Science Monthly* 77 (1910): 163-69.

23. "Bear Family Hate Solved in Display," *Los Angeles Times*, clipping date-stamped November 18, 1951.

24. D. Rodney Lee, "The Knudsen-Machris Expedition to Kenya," *Museum Quarterly* 4, no. 4 (1966): 15-20.

25. Letter from Museum Taxidermist J. P. Herring to *Outdoor Life* Managing Editor J. A. McGuire, January 2, 1920, main filing cabinet, box 8, folder 234, item 84, Tim Bovard Taxidermy Collection, NHM Archives and Special Collections.

26. Friedmann, "Artistry of the Habitat Group," 5.

27. Friedmann, 5.

28. Memorandum from Dr. Kenneth E. Stager to Assistant Director Gehring, November 22, 1972, NHM.ARC.008, Directors Files, box 3, folder 14, items 01-04, Giles W. Mead Collection, NHM Archives and Special Collections.

29. Hanson Puthuff, "Background," *Museum Graphic* 1, no. 2 (1926): 73-74.

30. George Adams, "Taxidermy and Sculpture—Keys to Natural History Exhibits," *Museum Quarterly* 4, no. 4 (1966): 21-26.

31. "McClellan Home with Trophies," *Los Angeles Times*, October 12, 1919.

32. Donna Haraway, "Teddy Bear Patriarchy: Taxidermy in the Garden of Eden, New York City, 1908-1936," *Social Text*, 11 (1985): 20-64.

33. "Americans Oppose Big Game Hunting . . . More than Six in Ten Favor Legal Ban," HBO Real Sports/Marist Poll, November 24, 2015. https://maristpoll.marist.edu/polls/1124-americans-oppose-big-game-hunting-more-than-six-in-ten-favor-legal-ban/.

34. Lincoln, "History of the Habitat Group," 24.

Contributors

Kayce C. Bell
Natural History Museum of Los Angeles County

Lori Bettison-Varga
Natural History Museum of Los Angeles County

Tim Bovard
Natural History Museum of Los Angeles County

Jon Christensen
University of California, Los Angeles

Matt Davis
Natural History Museum of Los Angeles County

Ursula K. Heise
University of California, Los Angeles

Karen L. Maness
University of Texas at Austin

Araceli Ramos
California State University, Long Beach

Allison J. Shultz
Natural History Museum of Los Angeles County

Photo Credits

© California Academy of Sciences: 39 (bottom right), N21372.

© Craig Cutler, photographer: back cover.

p. 22 (left): William T. Hornaday, *Our Vanishing Wildlife* (New York: Charles Scribner's Sons, 1913).

p. 30 (top): *Mad Men*, courtesy of Lions Gate Television.

© Natural History Museum of Los Angeles County: 1, NHAI1007041; 4, NHAI1008099; 10–11, NHAI1007148; 15, NHAI1007476; 20 (bottom), NHAI1005141; 21, NHAI1007476; 22 (right), NHAI1008313; 24, NHAI1008131; 26, NHAI1007404; 28 (top), Tanzania, NHAI1007224; 28 (bottom), NHAI1007654; 30 (bottom), NHAI1007819; 38, 39 (top left), NHAI1007999; 39 (bottom left), NHAI1007717; 40 (top), NHAI1008136; 40 (bottom), NHAI1007035; 41, NHAI1008213; 42, NHAI1007683; 43 (top), Australia, NHAI1007737; 43 (bottom), Selengai, Kenya, NHAI1007707; 44 (middle), Chadwick Mountain, British Columbia, Canada, NHAI1008068; 47 (top), NHAI1005399; 47 (bottom), NHAI1008260; 48, NHAI1008404; 53 (top), NHAI1008184; 53 (bottom), NHAI1000521; 55 (top), NHAI1007682; 56, NHAI1007204; 58 (top), NHAI1007429; 58 (middle), NHAI1007063; 59 (top), NHAI1008194; 59 (middle), NHAI1007569; 60 (top), NHAI1008229; 60 (bottom left), NHAI1008161; 60 (bottom right), NHAI1008189; 61 (top left), NHAI1007424; 61 (top right), NHAI1007552; 61 (middle), NHAI1005447; 64, 65 (top left), NHAI1007516; 65 (top right), Tanzania, NHAI1007983; 65 (bottom), NHAI1007970; 68, Chihuahua, Mexico, NHAI1000514; 76 (left), 76 (right), NHAI1007652; 77 (top), NHAI1008218; 77 (bottom), NHAI1008217; 80 (left), Fraser River Canyon, British Columbia, NHAI1008176; 81 (top), Effigy Mounds, Iowa, NHAI1008382; 82 (top), NHAI1008407; 82 (bottom), NHAI1007039; 83 (top), NHAI1008237; 84 (middle), NHAI1007381; 84 (bottom), NHAI1000529; 85, NHAI1007677; 86 (right), NHAI1007567; 91 (top), NHAI1007107; 94 (top), Mahale Mountains, Tanzania, NHAI1008332; 96 (top left), NHAI1007386; 96 (top right), Government Camp, Oregon, NHAI1008384; 97 (top left), Thika, Kenya, NHAI1008337; 97 (top middle), Lancaster County, Pennsylvania, NHAI1008342; 97 (top right), Guama, Venezuela, NHAI1008341; 97 (bottom), Congo Basin, NHAI1008397; 98 (top left), Ituri Forest, Democratic Republic of the Congo, NHAI1008393; 98 (bottom left), NHAI1008173; 99 (left), NHAI1008333; 99 (right), NHAI1008256; 102, NHAI1007285; 104 (top), NHAI1007230; 107 (left), Tana River, Kenya, NHAI1007627; 107 (top right), Chadwick Mountain, British Columbia, NHAI1008064; 107 (bottom right), NHAI1007555; 108 (top left), NHAI1007820; 108 (top right), NHAI1007853; 108 (bottom), NHAI1007854; 109 (top left), NHAI1007827; 109 (top right), NHAI1007838; 109 (bottom), NHAI1007830; 112 (left), NHAI1007534; 114, NHAI1008411; 116 (top), NHAI1008105; 116 (bottom), NHAI1008105; 117 (top left), NHAI1007389; 117 (top right), NHAI1007390; 117 (bottom), NHAI1007176; 118 (top), NHAI1007321; 118 (bottom left), NHAI1007078; 118 (bottom right), NHAI1007281; 119, NHAI1008413; 123 (bottom), NHAI1007036; 136, NHAI1007042; 138, 139 (left), NHAI1008307; 139 (right), Buena Vista Lake, California, NHAI1008116. **Tom Bonner, photographer:** 18. **Tim Bovard, photographer:** 110. **Chris Candelaria, photographer:** 122. **Gina Cholick, photographer:** 23, 28 (middle), 33 (top), 36, 120, 123 (top left), 123 (top right). **Deniz Durmus, photographer:** 6–7 (dedication), 14, 20 (top), 126. **Christina Gandolfo, photographer:** 104 (bottom). **Natalja Kent, photographer:** front cover, 2, 32, 44 (top), 44 (bottom), 45, 46, 49, 50, 54, 55 (bottom), 61 (bottom), 62, 66–67, 69 (bottom), 70, 71, 74, 78, 81 (bottom), 83 (bottom), 86 (left), 87, 90, 91 (bottom), 92, 94 (bottom), 95, 96 (bottom), 98 (top right), 98 (middle right), 98 (bottom right), 100, 101, 105, 106, 111, 112 (right), 113. **Mario de Lopez, photographer:** 8, 12, 16–17, 25 (top), 29, 31, 33 (bottom), 34–35, 57, 58 (bottom), 59 (bottom), 69 (top), 72–73, 80 (right), 84 (top), 88–89, 124. **Seaver Center for Western History Research:** endpapers, 1924 blueprints. **Zack Williams, photographer:** 25 (bottom).

p. 39 (top right), p. 52: John Rowley, *Taxidermy and Museum Exhibition* (New York: D. Appleton, 1925).

Index

This book is published in conjunction with the exhibition *Reframing Dioramas: The Art of Preserving Wilderness*, presented at the Natural History Museum of Los Angeles County from 2024 to 2025.

Reframing Dioramas: The Art of Preserving Wilderness is among more than 60 exhibitions and programs presented as part of PST ART. Returning in September 2024 with its latest edition, PST ART: *Art & Science Collide*, this landmark regional event explores the intersections of art and science, both past and present. PST ART is presented by Getty. For more information about PST ART: *Art & Science Collide*, please visit pst.art.

PST ART
ART & SCIENCE COLLIDE
Presented by
Getty

NATURAL HISTORY MUSEUM
LOS ANGELES COUNTY

Library of Congress Control Number: 2024931709
ISBN 978-1-64657-040-9

Published by the Natural History Museum of Los Angeles County
900 Exposition Boulevard
Los Angeles, CA 90007
www.nhm.org

Available through Artbook | D.A.P.
75 Broad Street, Suite 630
New York, NY 10004
www.artbook.com

Produced by Marquand Books, Seattle
www.marquandbooks.com

Edited by Beth Chapple
Designed by Ryan Polich
Typeset in Area Normal and Articulat CF by Brynn Warriner
Proofread by Janice Lee
Index by David Luljak
Color management by I/O Color, Seattle
Printed and bound in China by Artron Art Group

Page 1
Taxidermist Paul Hill adds finishing touches to a beetle in the first *Chimpanzee* diorama (1965).

Frontispiece
Version 2 of the *Grizzly Bear* diorama combined existing taxidermy with a new foreground and background (2021).

Title spread
The *Topi and Warthog* group around 1962 before the foreground was added to the diorama.

Dedication
Habitats shine out from the Level 2 Hall of North American Mammals like windows into different worlds (2016).

Page 8
The *Bongo* diorama reveals an intimate moment hidden among the bamboo groves of the Aberdare Mountains in Kenya (2014).

Back cover
Taxidermist Tim Bovard vacuums the elephants in the *Watering Hole* diorama to prepare the museum for reopening after Covid-19 pandemic restrictions were relaxed (2021).

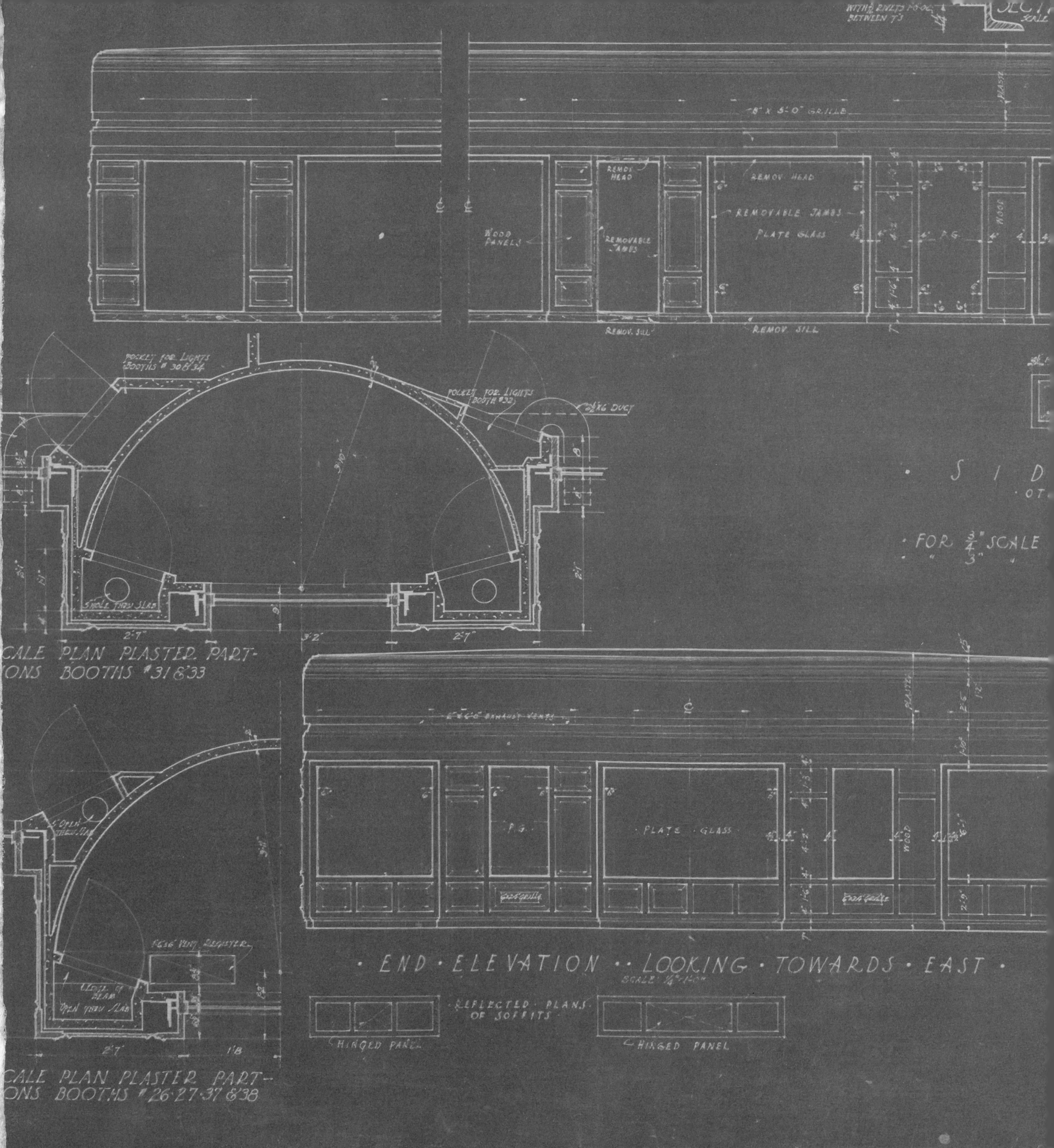

REMOV. HEAD
REMOVABLE JAMBS
PLATE GLASS
WOOD PANELS
REMOV. SILL
POCKET FOR LIGHTS BOOTHS # 30 & 34
POCKET FOR LIGHTS BOOTH #32
DUCT
S I D
FOR 3/4" SCALE
CALE PLAN PLASTER PART-
IONS BOOTHS #31 & 33
EXHAUST VENTS
P.G.
PLATE GLASS
WOOD
· END · ELEVATION · · LOOKING · TOWARDS · EAST ·
SCALE: 1/4"=1'-0"
REFLECTED PLANS OF SOFFITS
HINGED PANEL
HINGED PANEL
CALE PLAN PLASTER PART-
IONS BOOTHS #26·27·37 & 38

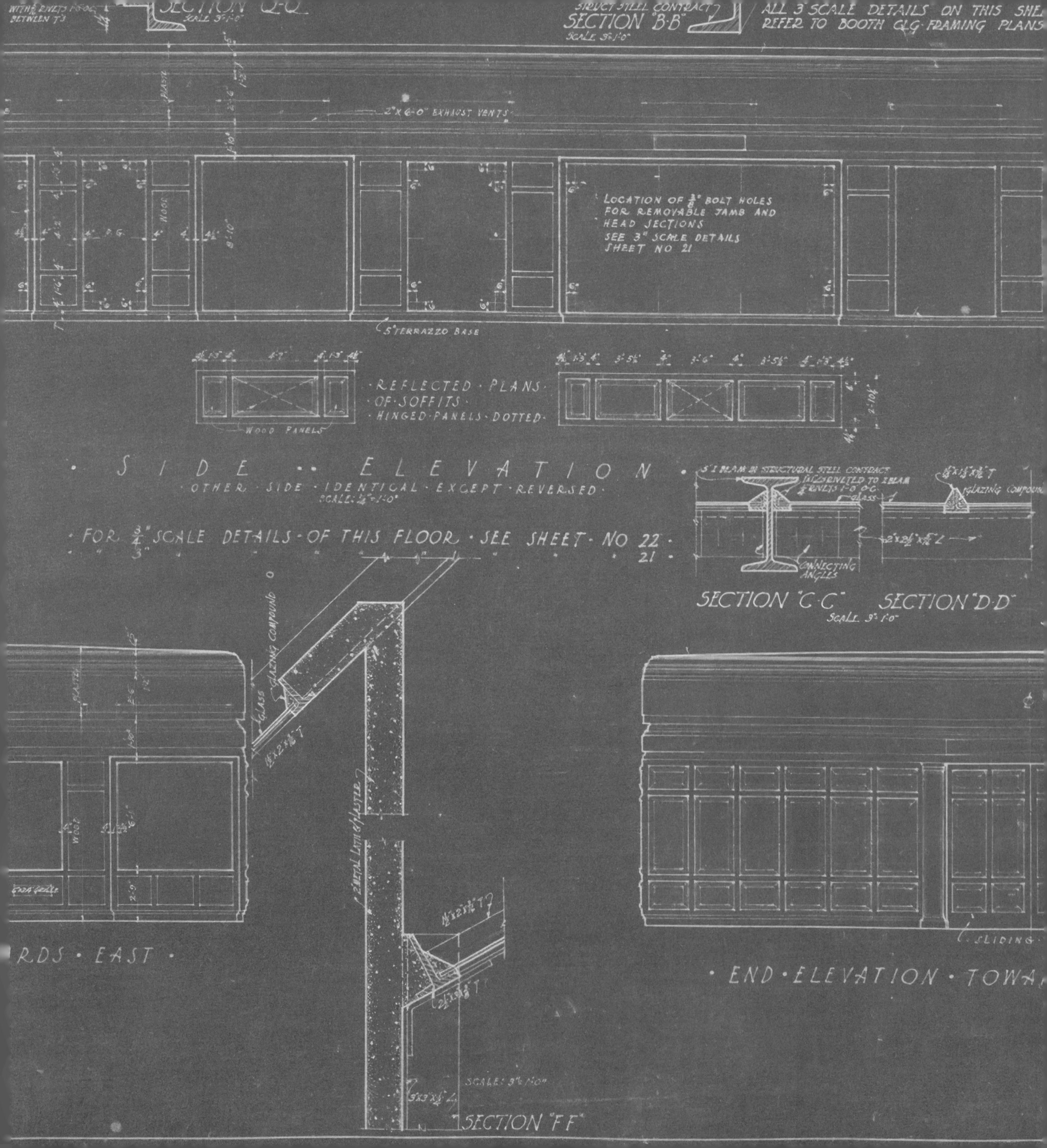

SECTION Q-Q
SCALE 3"=1'-0"
STRUCT STEEL CONTRACT
SECTION "B-B"
SCALE 3"=1'-0"
ALL 3" SCALE DETAILS ON THIS SHE
REFER TO BOOTH CLG FRAMING PLANS
2"X 6'-0" EXHAUST VENTS
LOCATION OF 3/8" BOLT HOLES
FOR REMOVABLE JAMB AND
HEAD SECTIONS
SEE 3" SCALE DETAILS
SHEET NO 21
5" TERRAZZO BASE
REFLECTED PLANS
OF SOFFITS
HINGED PANELS DOTTED
WOOD PANELS
SIDE ELEVATION
OTHER SIDE IDENTICAL EXCEPT REVERSED
SCALE: 1/4"=1'-0"
FOR 3/4" SCALE DETAILS OF THIS FLOOR SEE SHEET NO 22
" 3" " 21
5" I BEAM IN STRUCTURAL STEEL CONTRACT
GLASS
GLAZING COMPOUND
CONNECTING ANGLES
SECTION "C-C"
SECTION "D-D"
SCALE 3"=1'-0"
GLAZING COMPOUND
GLASS
METAL LATH & PLASTER
RDS EAST
SCALE: 3"=1'-0"
SECTION "F-F"
SLIDING
END ELEVATION TOWA